SpringerBriefs in Molecular Science

Biobased Polymers

Series editor

Patrick Navard, Centre de Mise en Forme des Matériaux, Ecole des Mines Paris, Mines ParisTech, Sophia Antipolis cedex, France

Published under the auspices of EPNOE*Springerbriefs in Biobased polymers covers all aspects of biobased polymer science, from the basis of this field starting from the living species in which they are synthetized (such as genetics, agronomy, plant biology) to the many applications they are used in (such as food, feed, engineering, construction, health, …) through to isolation and characterization, biosynthesis, biodegradation, chemical modifications, physical, chemical, mechanical and structural characterizations or biomimetic applications. All biobased polymers in all application sectors are welcome, either those produced in living species (like polysaccharides, proteins, lignin, …) or those that are rebuilt by chemists as in the case of many bioplastics.

Under the editorship of Patrick Navard and a panel of experts, the series will include contributions from many of the world's most authoritative biobased polymer scientists and professionals. Readers will gain an understanding of how given biobased polymers are made and what they can be used for. They will also be able to widen their knowledge and find new opportunities due to the multidisciplinary contributions.

This series is aimed at advanced undergraduates, academic and industrial researchers and professionals studying or using biobased polymers. Each brief will bear a general introduction enabling any reader to understand its topic.

*EPNOE The European Polysaccharide Network of Excellence (www.epnoe.eu) is a research and education network connecting academic, research institutions and companies focusing on polysaccharides and polysaccharide-related research and business.

More information about this series at http://www.springer.com/series/15056

Carla Vilela · Ricardo João Borges Pinto
Susana Pinto · Paula Marques · Armando Silvestre
Carmen Sofia da Rocha Freire Barros

Polysaccharide Based Hybrid Materials

Metals and Metal Oxides, Graphene and Carbon Nanotubes

 Springer

Carla Vilela
Department of Chemistry, CICECO—
 Aveiro Institute of Materials
University of Aveiro
Aveiro, Portugal

Ricardo João Borges Pinto
Department of Chemistry, CICECO—
 Aveiro Institute of Materials
University of Aveiro
Aveiro, Portugal

Susana Pinto
Mechanical Engineering Department,
 TEMA—Centre for Mechanical
 Technology and Automation
University of Aveiro
Aveiro, Portugal

Paula Marques
Mechanical Engineering Department,
 TEMA—Centre for Mechanical
 Technology and Automation
University of Aveiro
Aveiro, Portugal

Armando Silvestre
Department of Chemistry, CICECO—
 Aveiro Institute of Materials
University of Aveiro
Aveiro, Portugal

Carmen Sofia da Rocha Freire Barros
Department of Chemistry, CICECO—
 Aveiro Institute of Materials
University of Aveiro
Aveiro, Portugal

ISSN 2191-5407   ISSN 2191-5415 (electronic)
SpringerBriefs in Molecular Science
ISSN 2510-3407   ISSN 2510-3415 (electronic)
Biobased Polymers
ISBN 978-3-030-00346-3   ISBN 978-3-030-00347-0 (eBook)
https://doi.org/10.1007/978-3-030-00347-0

Library of Congress Control Number: 2018955915

This Springer imprint is published by the registered company Springer Nature Switzerland AG
The registered company address is: Gewerbestrasse 11, 6330 Cham, Switzerland

Preface

Polysaccharides, the most abundant family of natural polymers, had gained considerable attention in the last decades as a source of innovative bio-based materials, including an extensive assortment of polysaccharide hybrid nanomaterials for distinct applications. This book presents the current knowledge about polysaccharide-based hybrid nanomaterials with metal and metal oxide nanoparticles, carbon nanotubes and graphene. The book covers the main polysaccharides, namely cellulose, chitin, chitosan and starch, as well as their most relevant derivatives, and features the description of the most significant production methodologies, properties and utmost applications of these types of hybrids.

Keywords Polysaccharides · Hybrid materials · Metal nanoparticles
Graphene · Carbon nanotubes

Aveiro, Portugal

Carla Vilela
Ricardo João Borges Pinto
Susana Pinto
Paula Marques
Armando Silvestre
Carmen Sofia da Rocha Freire Barros

Acknowledgements

This work was developed within the scope of the project CICECO—Aveiro Institute of Materials (POCI-01-0145-FEDER-007679; UID/CTM/50011/2013) and TEMA (UID/EMS/00481/2013), financed by national funds through the FCT/MEC and when appropriate co-financed by FEDER under the PT2020 Partnership Agreement. The Portuguese Foundation for Science and Technology (FCT) is also acknowledged for the postdoctoral grants to R. J. B. Pinto (SFRH/BPD/89982/2012) and C. Vilela (SFRH/BPD/84168/2012), doctoral grant to S. Pinto (SFRH/BD/111515/2015) and research contracts under Investigador FCT to C. S. R. Freire (IF/01407/2012) and P. A. A. P. Marques (IF/00917/2013/CP1162/CT0016).

Contents

Chapter 1
Introduction

The quest to develop alternative eco-friendly materials derived from renewable resources to replace (partially or even totally) petroleum-based materials, is mainly devoted to the exploitation of naturally occurring polymers. In fact, natural polymers have gained the status of building-blocks to engineer multifunctional materials due to their abundance, low cost, biodegradability, biocompatibility and multiple functionalities [1–4]. A variety of natural polymers, such as polysaccharides [*e.g.*, cellulose, chitin, chitosan (CH), starch, alginate (ALG), dextran, fucoidan, heparin, hyaluronan and pullulan] and proteins (*e.g.*, albumin, apoferritin, casein, collagen, fibrinogen and gelatin), have been used for the development of all kinds of materials for the most assorted applications [5–8].

1.1 Polysaccharides

Polysaccharides are biopolymers composed of monosaccharides linked by glycosidic bonds. These biopolymers have different origins and sources, namely from plants (*e.g.*, cellulose and starch), animals (*e.g.*, chitin, heparin and hyaluronan), algae (*e.g.*, carrageenan, fucoidan and ALG) and microbial [*e.g.*, pullulan, dextran and bacterial cellulose (BC)], which in addition to biodegradability and biocompatibility exhibit diverse bioactivities, such as immunoregulatory, anti-tumour, anti-virus, anti-inflammatory, antioxidant and hypoglycemic activities [9]. Within the available polysaccharides, cellulose, chitin and its derivative CH, and starch (Fig. 1.1) are among the most studied biopolymers for the fabrication of a wide spectrum of functional materials. Despite the structural similarities between these polysaccharides (Fig. 1.1) with the main differences residing primarily on molecular weight (polymerization degree), the position and/or stereochemistry of the glycosidic bond as well as on the occurrence or not of branching, and ultimately, but of utmost importance on the functional group present at C2 in each saccharide unit (*i.e.*, OH group in cellulose and starch, $NHC(=O)CH_3$ in chitin, NH_2 in CH), their properties, namely crys-

© The Author(s), under exclusive license to Springer Nature Switzerland AG 2018
C. Vilela et al., *Polysaccharide Based Hybrid Materials*, Biobased Polymers,
https://doi.org/10.1007/978-3-030-00347-0_1

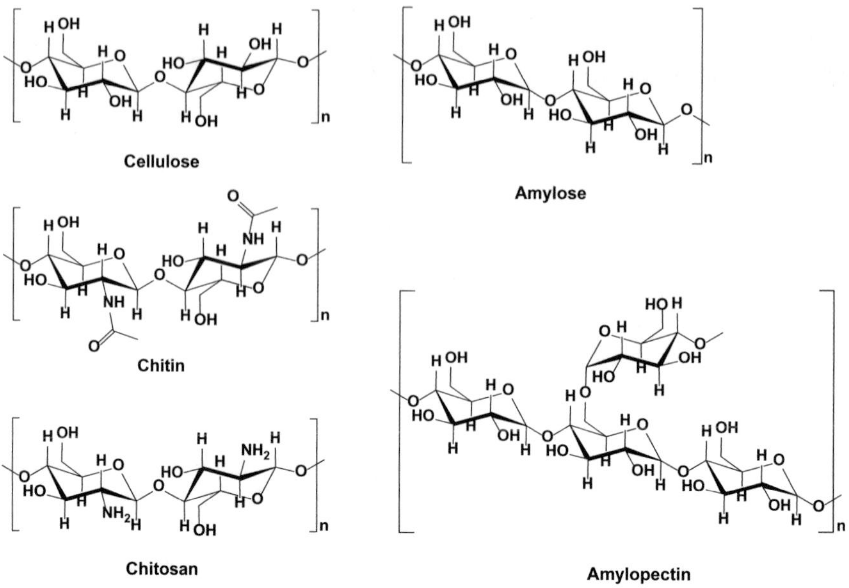

Fig. 1.1 Structure of cellulose, chitin, chitosan and starch (amylose and amylopectin)

tallinity, solubility and ability for chemical modification, are quite divergent. Other polysaccharides that are also at the spotlight include ALG, *i.e.* an anionic polysaccharide derived from seaweeds [10], hyaluronan, *i.e.* an anionic glycosaminoglycan available in vertebrate tissues [11, 12] and carrageenan, *i.e.* sulphated polysaccharide derived from red seaweeds [13]; nevertheless, only some examples will be given regarding these polysaccharides since they are not the focus of the present book.

Cellulose is a linear homopolysaccharide composed of β-D-glucopyranose units linked by β-(1,4) glycosidic bonds (Fig. 1.1). The clear majority of cellulose available on earth is produced by photosynthesis in green plants, where it represents the main component of plant cell walls, associated with lignin and hemicelluloses. Nevertheless, this natural polymer is also produced by a family of sea animals called tunicates, several species of algae and some aerobic non-pathogenic bacteria [14, 15]. The discovery of the nanoscale forms of this ubiquitous, biodegradable and inexpensive biopolymer, *i.e.* cellulose nanofibrils (CNFs), cellulose nanocrystals (CNCs) and bacterial cellulose (BC), unlocked novel perspectives for the design of sustainable nanomaterials for a multitude of applications [16]. The most relevant properties of this polysaccharide include anisotropic shape, excellent mechanical properties, good biocompatibility and tailorable surface chemistry. Further details regarding cellulose structure, properties and applications can be explored in the relevant literature [14, 15, 17–20].

Chitin is a high molecular weight linear homopolysaccharide consisting of *N*-acetyl-2-amido-2-deoxy-*D*-glucose units linked by β-(1,4) glycosidic bonds (Fig. 1.1). This polysaccharide is the second most abundant biopolymer and the main component of the exoskeleton of crustaceans, molluscs and insects [21]. At an industrial level chitin is easily obtained from the shells of crabs, shrimps and lobsters originated from the sea food processing waste shells. Despite the poor solubility and processability of chitin, the biodegradability and biocompatibility of this polysaccharide makes it an asset for biomedical applications [22].

CH is also a high molecular weight linear heteropolysaccharide obtained from chitin via N-deacetylation in different degrees [21]. It is mainly composed of 2-amino-2-deoxy-D-glucose units linked through β-(1,4) glycosidic bonds (Fig. 1.1). This polysaccharide has a cationic character and exhibits unique properties, such as biocompatibility, antimicrobial activity and excellent film-forming ability, which makes it particularly appealing for diverse applications [23, 24]. Supplementary details concerning the structure, properties and applications of chitin and CH are available elsewhere [21–27].

Starch is a naturally occurring storage heteropolysaccharide that consists of two macromolecules, namely amylose and amylopectin (Fig. 1.1), whose proportions vary with plant origin [28]. While amylose, a linear polysaccharide of glucose units linked through α-(1,4) glycosidic bonds, accounts for about 20–30% of starch composition, the amylopectin, a multi-branched macromolecular component with additional α-(1,6) linkages, accounts for *ca.* 70–80% of starch composition [29]. Starch can be found in a variety of plant organs such as cereal grains and tubers and is often described according to its origin as *e.g.*, corn starch, potato starch, tapioca starch, etc. [28]. Albeit the insolubility of starch in cold water and alcohols, this polysaccharide is soluble in hot water via a gelatinization process where water acts as a plasticizer. Comprehensive reviews about starch are also available elsewhere [28–31].

Polysaccharides sparked the imagination of scientists, who thus have been using them to create multifunctional materials for a multitude of applications, including food packaging [32], osteoarthritis therapy [33], vaccines [34], nanotherapeutics [1], drug delivery [35] and theranostics [36], among many others. In addition, this fascinating class of biopolymers are also being exploited for the development of functional hybrid materials for various domains spanning from biomedical to technological applications [37–39]. Just to highlight a few, cellulose was combined with quantum dots to design photoluminescence nanohybrids for anti-counterfeiting applications [40], the partnership between CH and silica originated hybrid porous membranes [41], ALG, CH and golden single-walled carbon nanotubes (SWCNTs) yielded an effective hybrid photothermal converter for cancer ablation [42], and chitin was combined with graphene oxide (GO) to fabricate hybrid materials for the removal of pollutant dyes [43].

1.2 Hybrid Materials

Hybrid materials comprise two or more constituents with different natures, *i.e.* at least one of the constituents is inorganic and the other is organic. The mixing and/or interaction between the constituents usually occurs at the micrometric and sub-micrometric scale, reaching down to the nanometric and molecular level [44]. The ensuing hybrid materials have either numerous functionalities and/or novel properties due to the interactions between the individual constituents, mostly associated with synergetic effects [45, 46]. Examples of nature made hybrid materials include nacre, which is a crystallized compacted lamellar structure composed of aragonite and conchiolin, and the natural pigment known as Blue Maya, which results from the combination of natural dyes (derived of indigo-type molecules) and lamellar clays [47].

Hybrid materials can be roughly divided into two distinct classes according to the nature of the interfacial interactions between the phases/components: (i) Class I, *i.e.* the organic and inorganic components are embedded, and the cohesion of the whole structure is due to hydrogen, van der Waals or electrostatic bonds, and (ii) Class II, *i.e.* strong chemical covalent bonds partially link together the distinct components [48]. These materials can be prepared by bottom-up strategies (Fig. 1.2) including those from molecular precursors and well-defined "nano-objects", as well as template-based strategies [49]. These methodologies can make use of processing approaches such as casting, electrospinning, dip-, spin- and spray-coating, soft/hard lithography and spray-drying, which can originate a multitude of materials like for example monoliths, foams, fibres, membranes, films, patterns and particles as depicted in Fig. 1.2 [49]. The main chemical routes for the design of functional hybrid nanomaterials do not require any extensive coverage here, given the published comprehensive reviews on the topic [46–52].

Organic-inorganic hybrid materials have attracted the interest of various researchers due to their unpaired mechanical, optical, electrical and thermal properties, which allow them to be applied in several domains such as mechanics, optics, electronics, energy, environment, biology and medicine [46, 50]. Manifold hybrid organic-inorganic materials with great potential for high added-value applications have been developed over the past five years, mainly due to the enormous flexibility of the synthetic routes and the almost endless choices of possible combinations that can be employed to fabricate organic-inorganic hybrid structures [44, 46, 47, 51, 53, 54].

Numerous interesting reviews about organic/inorganic hybrid materials containing polysaccharides have been published, like for example the reviews on polysaccharides/silica hybrids for biomedical and industrial applications [38], the efficient hybrid association between metal oxides and polysaccharides [37], hybrid hydrogels based on polysaccharides for cartilage regeneration [55], and hybrid systems of CNCs and inorganic nanoparticles with potential applications in biomedical and chemical systems [56]. Worth mentioning is also the chapter devoted to hybrid materials composed of polysaccharides (cellulose, chitin and CH) for biomedical applications such as biosensors, actuators, theranostics and tissue engineering [39].

As long as our literature survey could ascertain, there have been no comprehensive appraisals gathering the information that highlights the huge variety of polysaccharides-based hybrids for diverse domains of application. In this context, the purpose of this book is not to cover exhaustively the numerous publications dealing with polysaccharides-based hybrids, but rather to select representative studies published in the last 5 years that originated fascinating materials derived from cel-

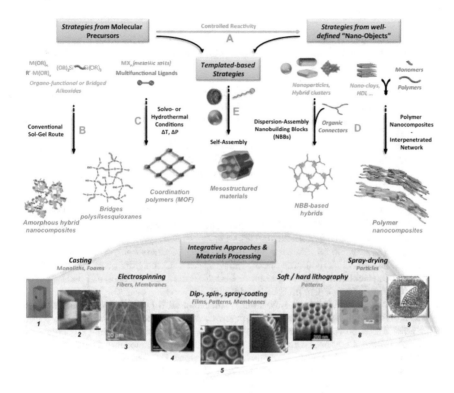

Fig. 1.2 The main routes for the design of functional hybrid nanomaterials, together with processing approaches and examples of resulting materials. Reprinted with permission from [49]. Copyright 2014 American Chemical Society

lulose, chitin, chitosan and starch, and containing metal nanoparticles [Au, Ag, Cu and Pd (Chap. 2)], metal oxide nanoparticles [TiO_2, ZnO, CuO, Cu_2O, SiO_2 Fe_2O_3 and Fe_3O_4 (Chap. 3)] and carbon nanomaterials [graphene (Chap. 4) and carbon nanotubes, CNTs (Chap. 5)].

References

1. Mizrahy S, Peer D. Polysaccharides as building blocks for nanotherapeutics. Chem Soc Rev. 2012;41:2623–40.
2. Yang X, Shi X, D'arcy R, Tirelli N, Zhai G. Amphiphilic polysaccharides as building blocks for self-assembled nanosystems: molecular design and application in cancer and inflammatory diseases. J Control Release. 2018;272:114–44.
3. Fuenzalida JP, Goycoolea FM. Polysaccharide-protein nanoassemblies: novel soft materials for biomedical and biotechnological applications. Curr Protein Pept Sci. 2015;16:89–99.
4. Srinivasan N, Kumar S. Ordered and disordered proteins as nanomaterial building blocks. WIREs Nanomed Nanobiotechnol. 2012;4:204–18.

5. Vilela C, Figueiredo ARP, Silvestre AJD, Freire CSR. Multilayered materials based on biopolymers as drug delivery systems. Expert Opin Drug Deliv. 2017;14:189–200.
6. Silva NHCS, Vilela C, Marrucho IM, Freire CSR, Pascoal Neto C, Silvestre AJD. Protein-based materials: from sources to innovative sustainable materials for biomedical applications. J Mater Chem B. 2014;2:3715–40.
7. Pinto RJB, Carlos LD, Marques PAAP, Silvestre AJD, Freire CSR. An overview of luminescent bio-based composites. J Appl Polym Sci. 2014;131:41169.
8. Song Y, Zheng Q. Ecomaterials based on food proteins and polysaccharides. Polym Rev. 2014;54:514–71.
9. Yu Y, Shen M, Song Q, Xie J. Biological activities and pharmaceutical applications of polysaccharide from natural resources: a review. Carbohydr Polym. 2018;183:91–101.
10. Lee KY, Mooney DJ. Alginate: properties and biomedical applications. Prog Polym Sci. 2012;37:106–26.
11. Dicker KT, Gurski LA, Pradhan-Bhatt S, Witt RL, Farach-Carson MC, Jia X. Hyaluronan: a simple polysaccharide with diverse biological functions. Acta Biomater. 2014;10:1558–70.
12. Dosio F, Arpicco S, Stella B, Fattal E. Hyaluronic acid for anticancer drug and nucleic acid delivery. Adv Drug Deliv Rev. 2015;97:204–36.
13. Campo VL, Kawano DF, Da Silva DB Jr, Carvalho I. Carrageenans: biological properties, chemical modifications and structural analysis—a review. Carbohydr Polym. 2009;77:167–80.
14. Klemm D, Heublein B, Fink H-P, Bohn A. Cellulose: fascinating biopolymer and sustainable raw material. Angew Chem Int Ed. 2005;44:3358–93.
15. Klemm D, Kramer F, Moritz S, Lindström T, Ankerfors M, Gray D, Dorris A. Nanocelluloses: a new family of nature-based materials. Angew Chem Int Ed. 2011;50:5438–66.
16. Abitbol T, Rivkin A, Cao Y, Nevo Y, Abraham E, Ben-Shalom T, Lapidot S, Shoseyov O. Nanocellulose, a tiny fiber with huge applications. Curr Opin Biotechnol. 2016;39:76–88.
17. Moon RJ, Martini A, Nairn J, Simonsen J, Youngblood J. Cellulose nanomaterials review: structure, properties and nanocomposites. Chem Soc Rev. 2011;40:3941–94.
18. Figueiredo ARP, Vilela C, Neto CP, Silvestre AJD, Freire CSR. Bacterial cellulose-based nanocomposites : roadmap for innovative materials. In: Thakur VK, editor. Nanocellulose polymer nanocomposites: fundamentals and applications. Scrivener Publishing LLC; 2014. p. 17–64.
19. Vilela C, Pinto RJB, Figueiredo ARP, Neto CP, Silvestre AJD, Freire CSR. Development and applications of cellulose nanofibers based polymer composites. In: Bafekrpour E, editor. Advanced composite materials: properties and applications. De Gruyter Open; 2017. p. 1–65.
20. Yang J, Li J. Self-assembled cellulose materials for biomedicine: a review. Carbohydr Polym. 2018;181:264–74.
21. Rinaudo M. Chitin and chitosan: properties and applications. Prog Polym Sci. 2006;31:603–32.
22. Anitha A, Sowmya S, Kumar PTS, Deepthi S, Chennazhi KP, Ehrlich H, Tsurkan M, Jayakumar R. Chitin and chitosan in selected biomedical applications. Prog Polym Sci. 2014;39:1644–67.
23. Patrulea V, Ostafe V, Borchard G, Jordan O. Chitosan as a starting material for wound healing applications. Eur J Pharm Biopharm. 2015;97:417–26.
24. Wang H, Qian J, Ding F. Emerging chitosan-based films for food packaging applications. J Agric Food Chem. 2018;66:395–413.
25. Ma J, Sahai Y. Chitosan biopolymer for fuel cell applications. Carbohydr Polym. 2013;92:955–75.
26. Ngo D-H, Vo T-S, Ngo D-N, Kang K-H, Je J-Y, Pham HN-D, Byun H-G, Kim S-K. Biological effects of chitosan and its derivatives. Food Hydrocolloids. 2015;51:200–16.
27. Choi C, Nam J-P, Nah J-W. Application of chitosan and chitosan derivatives as biomaterials. J Ind Eng Chem. 2016;33:1–10.
28. Alcázar-Alay SC, Meireles MAA. Physicochemical properties, modifications and applications of starches from different botanical sources. Food Sci Technol. 2015;35:215–36.
29. Zhu F. Structures, properties, and applications of lotus starches. Food Hydrocolloids. 2017;63:332–48.

30. Nafchi AM, Moradpour M, Saeidi M, Alias AK. Thermoplastic starches: properties, challenges, and prospects. Starch. 2013;65:61–72.
31. Masina N, Choonara YE, Kumar P, du Toit LC, Govender M, Indermun S, Pillay V. A review of the chemical modification techniques of starch. Carbohydr Polym. 2017;157:1226–36.
32. Cazón P, Velazquez G, Ramírez JA, Vázquez M. Polysaccharide-based films and coatings for food packaging: a review. Food Hydrocolloids. 2017;68:136–48.
33. Chen Q, Shao X, Ling P, Liu F, Han G, Wang F. Recent advances in polysaccharides for osteoarthritis therapy. Eur J Med Chem. 2017;139:926–35.
34. Cordeiro AS, Alonso MJ, de la Fuente M. Nanoengineering of vaccines using natural polysaccharides. Biotechnol Adv. 2015;33:1279–93.
35. Pushpamalar J, Veeramachineni AK, Owh C, Loh XJ. Biodegradable polysaccharides for controlled drug delivery. ChemPlusChem. 2016;81:504–14.
36. Liu Q, Duan B, Xu X, Zhang L. Progress in rigid polysaccharide-based nanocomposites with therapeutic functions. J Mater Chem B. 2017;5:5690–713.
37. Boury B, Plumejeau S. Metal oxides and polysaccharides: an efficient hybrid association for materials chemistry. Green Chem. 2015;17:72–88.
38. Salama A. Polysaccharides/silica hybrid materials: new perspectives for sustainable raw materials. J Carbohydr Chem. 2016;35:131–49.
39. Soares PIP, Echeverria C, Baptista AC, João CFC, Fernandes SN, Almeida APC, Silva JC, Godinho MH, Borges JP. Hybrid polysaccharide-based systems for biomedical applications. In: Thakur VK, Thakur MK, Pappu A, editors. Hybrid polymer composite materials applications. 1st ed. Woodhead Publishing, Elsevier Ltd.; 2017. p. 107–49.
40. Chen L, Lai C, Marchewka R, Berry RM, Tam KC. Use of CdS quantum dot-functionalized cellulose nanocrystal films for anti-counterfeiting applications. Nanoscale. 2016;8:13288–96.
41. Pandis C, Madeira S, Matos J, Kyritsis A, Mano JF, Ribelles JLG. Chitosan–silica hybrid porous membranes. Mater Sci Eng, C. 2014;42:553–61.
42. Meng L, Xia W, Liu L, Niu L, Lu Q. Golden single-walled carbon nanotubes prepared using double layer polysaccharides bridge for photothermal therapy. ACS Appl Mater Interfaces. 2014;6:4989–96.
43. González JA, Villanueva ME, Piehl LL, Copello GJ. Development of a chitin/graphene oxide hybrid composite for the removal of pollutant dyes: adsorption and desorption study. Chem Eng J. 2015;280:41–8.
44. Nicole L, Laberty-Robert C, Rozes L, Sanchez C. Hybrid materials science: a promised land for the integrative design of multifunctional materials. Nanoscale. 2014;6:6267–92.
45. Fahmi A, Pietsch T, Mendoza C, Cheval N. Functional hybrid materials. Mater Today. 2009;12:44–50.
46. Kickelbick G. Hybrid materials—Past, present and future. Hybrid Mater. 2014;1:39–51.
47. Díaz U, Corma A. Organic-inorganic hybrid materials: multi-functional solids for multi-step reaction processes. Chem – Eur J. 2018;24:3944–58.
48. Sanchez C, Rozes L, Ribot F, Laberty-Robert C, Grosso D, Sassoye C, Boissiere C, Nicole L. "Chimie douce": a land of opportunities for the designed construction of functional inorganic and hybrid organic-inorganic nanomaterials. C R Chim. 2010;13:3–39.
49. Sanchez C, Boissiere C, Cassaignon S, Chaneac C, Durupthy O, Faustini M, Grosso D, Laberty-Robert C, Nicole L, Portehault D, Ribot F, Rozes L, Sassoye C. Molecular engineering of functional inorganic and hybrid materials. Chem Mater. 2014;26:221–38.
50. Sanchez C, Julián B, Belleville P, Popall M. Applications of hybrid organic–inorganic nanocomposites. J Mater Chem. 2005;15:3559–92.
51. Hood MA, Mari M, Muñoz-Espí R. Synthetic strategies in the preparation of polymer/inorganic hybrid nanoparticles. Materials. 2014;7:4057–87.
52. Brendlé J. Organic–inorganic hybrids having a talc-like structure as suitable hosts to guest a wide range of species. Dalton Trans. 2018;47:2925–32.
53. Draxl C, Nabok D, Hannewald K. Organic/inorganic hybrid materials: challenges for ab initio methodology. Acc Chem Res. 2014;47:3225–32.

54. Cui J, Jia S. Organic–inorganic hybrid nanoflowers: a novel host platform for immobilizing biomolecules. Coord Chem Rev. 2017;352:249–63.
55. Sánchez-Téllez DA, Téllez-Jurado L, Rodríguez-Lorenzo LM. Hydrogels for cartilage regeneration, from polysaccharides to hybrids. Polymers. 2017;9:671.
56. Islam S, Chen L, Sisler J, Tam KC. Cellulose nanocrystal (CNC)—Inorganic hybrid systems: synthesis, properties and applications. J Mater Chem B. 2018;6:864–83.

Chapter 2
Polysaccharides-Based Hybrids with Metal Nanoparticles

For over a century, metallic nanoparticles (mNPs) have fascinated scientists, however, they have been empirically used by man since the middle ages as decorative pigments for colouring glass, like for example the famous Lycurgus Cup [1]. In this case, its use was due to one of the most interesting aspects of metallic colloids, their optical properties. The colour variation occurs due to the change in the surface plasmon resonance frequency (SPR) that is dependent on the size, and morphology but also on the refractive index of dispersant medium and the distance between adjacent mNPs [2].

In ancient times, the use of mNPs was certainly unintentional but, from a scientific point of view, the preparation of mNPs dates to the 19th century when Faraday described the preparation of monodisperse gold colloids by reduction of $[AuCl_4]^-$ ions using phosphorus in CS_2 as reducing agent [3]. Michael Faraday was probably the first scientist to correctly attribute the unusual properties of colloidal gold to size effects occurring on very small particle size [3]. Since this pioneering work, massive progresses have been made in the synthesis, functionalization and characterization of these nanosystems. The paramount goal of this research field is the control of the structure, size, shape and composition of the mNPs, since their properties are crucial to determine their functionality and, consequently, the final applicability [4].

Nowadays, it is well-known that mNPs exhibit distinct and, in some cases, unique physical, chemical and biological properties in comparison with their bulk counterparts. This justifies the enormous interest in these nanostructures in varied fields such as catalysis, electronics, sensors, medicine, among others [5].

The research on hybrid materials based in polysaccharides is a fast-growing area in materials science and engineering field. This growth results from the numerous potential applications that those materials can find [6]. The combination between polysaccharides and mNPs also contributed for the growth of this field where the "green" connotation and renewable nature of polysaccharides was mixed with novel and specific functionalities imparted by these inorganic fillers. This allows opening complete new areas of application until then impossible to achieve. Thus, considering the enormous development in this field, it is important to summarize the most

C. Vilela et al., *Polysaccharide Based Hybrid Materials*, Biobased Polymers,
https://doi.org/10.1007/978-3-030-00347-0_2

pertinent research results concerning the synthesis, properties and practical applications of these hybrid materials. In this vein, this chapter outlines distinct combinations between polysaccharides (cellulose, chitin, chitosan (CH) and starch) and mNPs, as well as the preparative methodologies and the potential applications of resulting hybrid materials. Gold (Au), silver (Ag), copper (Cu), and palladium (Pd) are the mNPs focused in this chapter.

2.1   Cellulose/mNPs Hybrid Materials

Since cellulose is the most abundant natural polymer is not surprising that this polysaccharide presents a huge window of opportunity when combined with distinct mNPs. Currently, it is observed a growing interest not only in the use of plant cellulose and its derivatives, but also the respective nanometric forms, namely cellulose nanofibrils (CNFs), cellulose nanocrystals (CNCs) and bacterial cellulose (BC), for the preparation of hybrid materials. The high number of publications in this domain validates the potentiality of cellulose-based hybrid materials, where the respective application varies depending on the mNPs type and the cellulose form. Several reviews in this field can be accessed to obtain a more detailed information on this type of materials. Some examples include the reviews by Pinto et al. [7] describing hybrid materials based in plant and bacterial cellulose, Foresti et al. [8] showing actual applications of BC/mNPs materials, and Islam et al. [6] reporting distinct CNCs/inorganic hybrid systems. Other appraisals are focused on specific applications of these hybrid materials like for example in catalysis [9], biosensing [10] and biomedical applications [11]. Table 2.1 summarizes some of the most recent examples of cellulose/mNPs hybrid materials reported in literature, as well as the respective preparation methodologies and practical applications.

Starting with cellulose and AuNPs, both have exciting features and their combination originates functional materials with unique properties. This topic was recently reviewed in detail by Van Rie and Thielemans [12] with emphasis on functional materials with specific catalytic, antimicrobial, sensing, antioxidant and Surface Enhanced Raman Scattering (SERS) performance. The application field of cellulose/AuNPs hybrids is broad but mainly centred on catalysis [13, 14] and sensors [15]. Nevertheless, other interesting applications include biomolecules recognition [16], and the design of optical [17, 18], conductive [19, 20], antioxidant [21], and antimicrobial [22] materials.

A great number of cellulose/AuNPs hybrid materials were used as catalysts for the reduction of 4-nitrophenol to 4-aminophenol, usualy using $NaBH_4$ as reducing agent. An example was reported by Chen et al. [13], where 2,2,6,6-tetramethylpiperidine-1-oxyl (TEMPO)-oxidized BC nanofibres were used as matrix (Fig. 2.1). Oxidized-BC/AuNPs nanohybrids containing AuNPs with an average diameter of 4.30 nm, showed superior catalytic properties than the unsupported AuNPs with a pseudo-first order rate constant of 6.75×10^{-3} s^{-1}, which is nearly 20 times faster. Moreover, the

Table 2.1 Examples of hybrid materials based on cellulose and mNPs, the preparation methodologies and potential applications

Metal NPs	Cellulose matrix	Methodology	Application	References
Au	TEMPO-oxidized BC	In situ reduction using $NaBH_4$	Catalysis (reduction of 4-nitrophenol)	[13]
	CNCs	Hydrothermal synthesis	Catalysis (reduction of 4-nitrophenol)	[14]
	Cellulose ester	In situ reduction using trisodium citrate	Sensors (determination of iodide ions)	[15]
	Cellulose paper	Covalent immobilization of pre-synthesized AuNPs	Biomedical (biorecognition)	[16]
	Cellulose nanorods	Direct mixing of the components	Optical (chiral plasmonic films)	[17]
	CNCs	Electrostatic binding of AuNPs	Optical (chiral plasmonics)	[18]
	CMC	Polymerization of aniline using $HAuCl_4$ as an oxidant	Thermal conductivity	[19]
	BC	In situ reduction using H_2O_2	Electrical conductivity	[20]
	Unbleached cellulose	In situ reduction in an autoclave	Antioxidant food packaging	[21]
	BC	Solution impregnation method	Biomedical (wound dressing)	[22]
Ag	BC	In situ reduction	Biomedical (wound dressing)	[23]
	CNCs	In situ reduction using *Syzygium cumini* leaf extract	Biomedical (wound healing)	[24]
	Cellulose acetate	Electrospinning	Water remediation (treatment of dye wastewater)	[25]
	Cellulose filter paper	In situ reduction using $NaBH_4$	Water remediation (clean drinking water)	[26]
	CNCs	Poly(dopamine) assisted reduction	Catalysis (reduction of 4-nitrophenol)	[27]
	CNFs	In situ reduction using $NaBH_4$	Sensors (detection of pesticides)	[28]
	MFC	Pyrrole adsorbed MFC aerogels dipped into $AgNO_3$ solution	Electrical conduction	[29]
	CMC	In situ reduction using natural honey	Anti-corrosion (Corrosion Inhibitor for St37 Steel)	[30]

(continued)

Table 2.1 (continued)

Metal NPs	Cellulose matrix	Methodology	Application	References
Cu	CNFs	In situ reduction using NaBH$_4$	Biomedical	[31]
	Regenerated cellulose	In situ reduction using *Ocimum sanctum* leaf extract	Antibacterial packaging and medical applications	[32]
	CMC	In situ reduction using hydrazine hydrate	Biomedical (urinary tract infection)	[33]
	Cotton fibres	In situ reduction using NaBH$_4$	Biomedical (wound dressing)	[34]
	BC	Magnetron sputtering	Electromagnetic shielding	[35]
	Cellulose fibres	In situ reduction using hydrazine hydrate	Catalysis (nitrodecarboxylation of aromatic unsaturated compounds)	[36]
Pd	CNCs	Reduction of Pd salt by the reducing ends of CNCs	Catalysis (Mizoroki–Heck cross-coupling reaction)	[37]
	CMC	In situ reduction using NaOH	Catalysis (degradation of azo-dyes)	[38]
	Ethylenediamine-functionalized cellulose	In situ reduction using NaBH$_4$	Catalysis (electrooxidation of hydrazine)	[39]
	Bio-waste corn-cob cellulose	In situ reduction using hydrazine hydrate	Catalysis (Suzuki-Miyaura cross-coupling reactions)	[40]

catalytic properties of these nanohybrids are dependent on the amount of NaBH$_4$, as well as on the temperature of the reaction mixture.

A different method for the development of new biosensor membranes was described by Li et al. [15], namely an effective method for the determination of iodide (I$^-$) ions by AuNPs deposited on a cellulose ester membrane. The authors showed that this hybrid material can be used as substrate in pulsed laser desorption/ionization mass spectrometry of high-salinity real samples such as edible salt samples and urine. This substrate presents a reduced background noise resulting from the binding of I$^-$ ions to the AuNPs, which induces an enhancement of the respective desorption and ionization efficiency. The high homogeneous nature of this hybrid probe improved the shot-to-shot and sample-to-sample reproducibility, thus enabling high accuracy in the measurements.

Majoinem et al. [18] exploited the twisting shape of CNCs as templates to prepare chiral plasmonics by binding cationic AuNPs on this negatively charged cellulosic

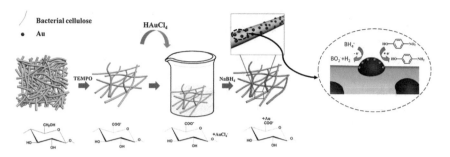

Fig. 2.1 Synthetic procedure for the preparation of oxidized-BC/AuNPs nanohybrid material and the correspondent illustration of its use in the catalytic reduction of 4-nitrophenol to 4-aminophenol using NaBH$_4$ aqueous solution. Adapted with permission from [13]. Copyright 2016 Elsevier

nanostructure. The electrostatic self-assembly leads to nanoscale fibrillar superstructures with lateral dimensions of 30–60 nm and length of 200–500 nm that exhibited a pronounced chiral right-handed plasmonic response, opposite to the left-handed of their liquid crystallinity assemblies. The authors verified that the sizes of CNCs and AuNPs must mutually match since too large AuNPs do not effectively bind on CNCs, and too small AuNPs do not provide strong enough plasmonic signal.

Interesting conducting hybrid materials using BC [20] or carboxymethyl cellulose [19] and AuNPs were also produced. In both cases, the cellulosic matrices were combined with poly(aniline) and then the AuNPs were adsorbed onto the polysaccharidic membranes. The presence of AuNPs induces differences in the voltammetric profile of the membranes leading to easier diffusional processes through the material. This allows an increased electrical conductivity and charge distribution over the membranes.

Li et al. [22] also showed the possibility of preparing BC/AuNPs based materials for treating bacterially infected wounds. BC membranes were soaked with colloidal AuNPs modified with 4,6-diamino-2-pyrimidinethiol with distinct concentrations. Modified hybrid membranes showed excellent physicochemical properties including water-uptake capability, mechanical strain, biocompatibility and better efficacy than most of current antibiotics (cefazolin/sulfamethoxazole) against Gram-negative bacteria, namely *Escherichia coli* and *Pseudomonas aeruginosa*. This was demonstrated on dorsal skin wounds of rats, where these hybrid materials inhibit bacterial growth and promote visible wound repair in 14 days.

Cellulose/AgNPs hybrids have also been extensively used in the development of antimicrobial materials for healthcare applications, such as wound healing [23, 24], because of the well known antimicrobial activity of AgNPs typically associated with the continous release of silver cations to the media. As representative examples, Wu et al. [23] used BC and AgNPs to prepare slow-released antimicrobial wound dressing hybrid materials. Uniform spherical AgNPs (10–30 nm) were generated and self-assembled homogeneously on the surface of BC nanofibres. This hybrid nanostructure offered excellent and sustainable control of Ag$^+$ release of

16.5% after 72 h in PBS solution. Regardless of the slow Ag^+ release, the hybrids exhibited significant antibacterial activity (more than 99% reduction against *E. coli* and *Staphylococcus aureus*, as well as *P. aeruginosa*) and, in co-culture with epidermal cells, did not showed cytotoxicity. These results manifested that BC/AgNPs gel-membrane hybrids were promising for antimicrobial wound dressing that could reduce inflammation and promote wound healing.

Cellulose/AgNPs hybrids are also used in other applications such as environmental protection (*e.g.* water remediation) [25, 26], catalysis [27], sensors [28], and the preparation of electrical conducting [29] and anti-corrosion [30] materials. For instance, Wang et al. [25] prepared cellulose acetate nanofibrous membranes with AgNPs, by electrospinning, for the treatment of dyes contaminated wastewater. The morphology and structures of the hybrid membranes can be controlled using solvent systems with distinct volatilities. The use of a solvent with a higher volatility results in a more porous structure with a higher ratio of ribbon-like fibres, offering a better dye (rhodamine B) adsorption ability that is not affected by the presence of AgNPs. As expected, the nanofibrous membranes containing AgNPs presented also an effective antibacterial activity against Gram-positive *S. aureus* and Gram-negative *E. coli* [25].

Another application of cellulose/AgNPs hybrids, in line with the global environmental concerns was reported by Liou et al. [28] on the detection of pesticides in fruits. Flexible and environmentally friendly substrates were prepared by impregnation of CNFs films with AgNPs and used in SERS analysis to detect thiabendazole (TBZ), a common pesticide and fungicide widely used for post-harvest tretament of fruits and vegetables. TBZ only exhibited strong SERS signals when the pH was below the TBZ's pKa value (pH = 4.65) and thus enabling the electrostatic attraction between TBZ and AgNPs. At a low pH value, CNFs prevented the uncontrolled aggregation of AgNPs serving as an effective platform for SERS analysis. The authors expected that this methodology could be extended in the future to the rapid detection of other neutral molecules and pesticides in various food products [28].

An interesting approach for the preparation of a benign corrosion inhibitor was studied by Solomon et al. [30] using carboxymethyl cellulose (CMC) as matrix and AgNPs produced in situ by reduction of $AgNO_3$ using honey as the capping and reducing agent. Hybrid materials with enhanced corrosion inhibitive ability for longer immersion times (until 15 h) in 15% H_2SO_4 solution, specifically for St37 steel, were prepared showing a high efficiency even at high temperatures (60 °C). The incorporation of AgNPs into CMC membranes increased the inhibition efficiency and stability of the polymer at high temperature and may also prevented the coiling up of the polymer under the same temperature conditions [30].

Cellulosic substrates can also be combined with CuNPs and the corresponding hybrid materials present similar applications to those of AgNPs-based hybrids, despite their fabrication and correspondent use being less significant. The cellulose/CuNPs hybrids can be obtained in the form of films [31, 32], hydrogels [33] or even as textile fibres [34] and are essentially used for as antimicrobial materials for biomedical applications. Following a physical approach, Lv et al. [35] used magnetron sputtering to prepare BC/CuNPs based hybrid materials with enhanced electromagnetic shielding, thermal, conduction, and mechanical properties. These topological constructed materials showed high conductivity (0.026 S m^{-1}), good mechanical properties (tensile strength ≈ 41.4 MPa and Young's modulus $= 7.02$ MPa) and acceptable interference (EMI) shielding effectiveness (55 dB). It is also important to refer that the use of BC/CuNPs materials in catalytic applications is also a possibility [36].

The combination of cellulose and PdNPs is also gaining prominence for the design of innovative catalytic materials. Some reports in this field describe the use of distinct types of cellulosic matrices such as CNCs [37], CMC [38], ethylenediamine-functionalized cellulose [39], and plant cellulose [40], as efficient matrix supports for PdNPs. Usually, the materials are prepared by in situ growth of PdNPs on the surface of the cellulosic materials, with the reducing ends of the matrix working as the reducing agents [37, 38] or through the addition of external reducing agents [39, 40]. These materials can be used for the degradation of azo-dyes in the presence of NaBH$_4$ [38] or for the electrooxidation of hydrazine [39].

Another example was given by Rezayat et al. [37] which prepared effective catalysts for carbon-carbon bond formation in the Mizoroki–Heck cross-coupling reaction. In this work, PdNPs were formed at the CNCs surface, where the reducing ends of CNCs acted both as the reducing agent and support material, using subcritical and supercritical CO$_2$ in one step for the easy recovery of the supported catalyst by simply venting with CO$_2$. The authors verified that the pressure, reaction time, and weight ratio of precursors control the palladium particle diameter and loading. After optimization of the experimental conditions, the mean diameter of PdNPs can varied between 6 and 13 nm, while the maximum Pd loading obtained was 45% (w/w) [37].

2.2 Chitin/mNPs Hybrid Materials

Although chitin is the second most abundant natural polymer, the number of reports dealing with chitin/mNPs based hybrid materials is scarce. The main factor that justifies this fact is most certainly the limited processability of this polysaccharide, which affects the simplicity and consequently the viability of the processes involved in the preparation of chitin-based materials. However, in last couple of years, some studies reported the combination of this biopolymer with AuNPs and AgNPs for the development of sensors [41] and materials for biomedical applications [42–44].

Huang et al. [41] reported the only work that focuses on the use of chitin nanofibrils (prepared via green physical method, namely a dilute acid facilitated cationization) as an efficient substrate to generate and immobilize AuNPs in a one-step process. The surface amino groups of chitin (with a degree of acetylation of 80.2%) acted as both reducing and stabilizing agents for the in situ synthesis of AuNPs due to the reducibility and chelation capacity of these groups. The size of the AuNPs (7–30 nm) were tuneable by adjusting the polysaccharide concentration, reaction time and temperature. The hybrid membrane exhibited a peroxidase mimic behaviour and, when combined with glucose oxidase, could be applied in the colourimetric detection of glucose with a detection limit of 94.5 nM [41].

Chitin/AgNPs hybrid materials also take advantage of the antimicrobial activity of AgNPs and can find practical applications as wound dressing [42], or as antifungal materials [43, 44]. These hybrids can be processed in distinct forms namely as powder [43, 44] or as membranes [42]. In the later example, Singh et al. [42] obtained chitin dressing cast films with AgNPs, which were synthesised by gamma irradiation at doses of 50 kGy and exhibited a particle size distribution in the range of 3–13 nm. In vitro antimicrobial tests showed that the membranes containing 100 ppm of AgNPs could completely inactivate viable *P. aeruginosa* cells within 1 h, while, for the same period, *S. aureus* was reduced by nearly 2-log CFU units.

2.3 Chitosan/mNPs Hybrid Materials

The versatility of CH as building block to engineer functional hybrid materials is well-known [45]. This polysaccharide, apart from its intrinsic biocompatibility, biodegradability and antimicrobial activity, presents a cationic behaviour in acidic solutions, film-forming ability, easy chemical or physical modifying ability, and strong affinity for metals ions [45]. These interesting properties turned this substrate into an excellent alternative to develop functional hybrid materials, particularly in combination with mNPs [46, 47]. Some relevant reviews in this area highlight distinct applications of CH/mNPs hybrids, namely on drug delivery [47], antimicrobial and wound healing [48] and catalysis [49]. Table 2.2 shows an overview of distinct CH-based hybrid materials with AuNPs, AgNPs, CuNPs and PdNPs, as well as the corresponding preparation methodologies and fields of applicability.

The combination of CH with AuNPs resulted, mainly, in the development of materials for sensors applications. The array of systems described in literature is massive, including sensors for the detection of glucose, uric acid, distinct metal ions (Cu^{2+}, Pb^{2+}, Hg^{2+}, Cd^{2+}), bacteria (*Bacillus cereus*, *E. coli*, *Salmonella typhimurium*), proteins (lectin), nucleic acids, antibiotics, virus, neurotransmitters (dopamine, tryptamine), pesticides, toxins, among others [46–49].

Dervisevic et al. [50] reported a sensor for the electrochemical detection of the adenosine-3-phosphate degradation product, xanthine, that can be successfully employed in the evaluation of meat freshness. Real samples of fish, chicken and beef were analysed for 25 days and, although no visual change was observed in the food condition, it was possible to detect an increase of the xanthine concentration. The use of self-assembled AuNPs resulted in an improved performance, namely in a lower response time (*ca.* 8 s), sensitivity (1.4 nA/μM), broader linear range (1–200 μM), and lower detection limit (0.25 mM) when compared to other studies using xanthine-based biosensors.

The development of food quality and safety control materials was also envisaged by Güner et al. [51] by targeting the specific detection of *E. coli*. In this work, a disposable hybrid film composed of CH, AuNPs, polypyrrole and carbon nanotubes

Table 2.2 Examples of hybrid materials based on CH and mNPs, the preparation methodologies and potential applications

Metal NPs	Methodology	Application	References
Au	Electrode modification using CH/polypyrrole/AuNPs hybrid	Sensors (meat quality)	[50]
	Electrode modification using CH/polypyrrole/AuNPs/CNTs hybrid film	Sensors (bacteria detection)	[51]
	Reduction of Au salt using CH	Biomedical (imaging agents)	[52]
	Reduction of Au salt using CH	Biomedical (drug delivery)	[53]
	Electrochemical reduction of gold and CH solution	Biomedical (orthopaedic implants)	[54]
	Blending of the components	Biomedical (scaffolds)	[55]
	Oil-in-water emulsion technique	Biomedical (nanotheranostics)	[56]
Ag	Blending of the components	Biomedical (antibacterial materials)	[57]
	Electrospun CH/poly(ethylene oxide) membranes with AgNPs	Biomedical (antibacterial membranes for tissue regeneration)	[58]
	Seed-mediated growth in presence of CH	Biomedical (non-invasive imaging)	[59]
	Coating of as-prepared AgNPs with quaternised CH	Sensors (detection of food contaminants)	[60]
	Sunlight-induced reduction of silver ions at CH microspheres surface	Environmental (water purification)	[61]
	Blending of the components	Environmental (removal of metals from surface waters)	[62]
	Emulsion-chemical cross-linking method followed by in situ deposition of AgNPs	Catalysis (4-nitrophenol reduction)	[63]
Cu	Blending of the components	Catalysis (C-S coupling reactions)	[64]
	Reduction of copper salt by L-ascorbic acid in presence of CH	Antifouling materials	[65]
	Sorption of copper salt in CH membranes followed by reduction with $NaBH_4$	Environmental (removal of metal ions from aquatic environment)	[66]
	Blending of the components	Agriculture (plant development and growth)	[67]
Pd	Reduction of Pd salt with ellagic acid in presence of modified CH	Catalysis (Suzuki–Miyaura C–C coupling reactions)	[68]
	Reduction of Pd salt with hydrazine in presence of modified CH	Catalysis (Suzuki–Miyaura C–C coupling reactions)	[69]

(continued)

Table 2.2 (continued)

Metal NPs	Methodology	Application	References
	In situ reduction of Pd salt in presence of montmorillonite/CH matrix	Catalysis (organic coupling reactions)	[70]
	Blending of the components	Biomedical (nanotheranostics)	[71]

(CNTs) was used to modify a pencil graphite electrode to construct an electrochemical immunosensor. In this platform, AuNPs allow the anti-*E. coli* monoclonal antibody immobilization and CNTs provide higher surface area on the electrode for antibodies accommodation. This biosensor presented a detection range from 30 to 3 × 10^7 CFU mL^{-1} with a detection limit of *ca.* 30 CFU mL^{-1} in PBS buffer. A pointed disadvantage of this hybrid material was its non-reusability since the antibodies were not able to regenerate for subsequent detections [51].

CH/AuNPs hybrids also found application in the development of materials for biomedical applications, such as imaging agents [52], drug delivery systems [53], orthopedic implants [54] and scaffolds [55]. For example, Tentor et al. [55] developed a scaffold for MC3T3-E1 osteoblast cells based on a CH/pectin hybrid loaded with AuNPs. This thermosensitive hydrogel promoted mild cell proliferation and growth over 10 days of exposure showing high cytocompatibility with several cell types, including normal kidney epithelial cells (VERO cells), epithelial colorectal adenocarcinoma cells (HT-29 cells), HPV-16 positive human cervical tumour cells (SiHa cells), kidney epithelial cells (LLCMK2 cells) and murine macrophage cells (J774A1 cells). Another interesting example was given by Kostevsek et al. [56] regarding a CH-based nanotheranostic system. Dumbbell-like gold–iron oxide NPs, prepared by a high temperature polyol method, were surface functionalized with modified-CH (thiol groups and cathecol fragments) via an oil-in-water emulsion technique and used as water-stable nanocarriers which proved to be effective as non-invasive photoacoustic imaging agent and as a photothermal therapy material [56].

In the case of CH/AgNPs hybrids, and as highlighted for the cellulose analogues (*i.e.* cellulose/AgNPs hybrids), most of the studies take advantage of the intrinsic antimicrobial activity of AgNPs, but here this feature is enhanced by the antimicrobial activity of CH [57, 58]. For instance, Holubnycha et al. [57] reported the preparation of CH/AgNPs hybrids with different component ratios that were tested against methicillin-resistant strains of *S. aureus* isolated from patients using a broth macro-dilution method. The authors modified the surface of AgNPs with cetrimonium bromide to improve their dispersibility and activity, as well as to decrease their toxicity. This hybrid showed a superior antimicrobial efficacy when compared to the pristine components, thus being a promising material to fight drug-resistant bacteria. Shao et al. [58] evaluated the biological activity of CH/AgNPs membranes prepared by electrospinning. The incorporation of AgNPs provided a continued antibacterial support to CH-based membranes in a dose-dependent manner. The in vitro and in vivo (subcutaneous implantation in rabbits) studies showed that AgNPs did not cause a noticeable cytotoxic effect on periodontal ligament cells. Furthermore, the membranes with AgNPs induced a similar inflammatory response compared with CH membranes without AgNPs, showing to be a promising material for clinical applications like guided tissue regeneration [58].

The CH/AgNPs hybrids can also present other interesting applications such as tracking and imaging [59], sensors for food contaminants [60], environmental (removal of dyes and microbial contaminants) [61], sorbents (extraction of metal pollutants from surface waters) [62], and catalytic [63] materials. For example, Chen et al. [60] used CH capped AgNPs to develop a highly sensitive detection system for food contaminants, namely tricyclazole and Sudan I, which are very difficult to detect at trace levels. Spherical functionalized AgNPs with sizes in the range 15–25 nm were prepared by microwave irradiation using quaternized CH since not only keeps the outstanding characteristics of CH but also displays great water solubility over a wide range of pH values. This hybrid system was then used to detect the target contaminants by SERS, presenting a limit of detection of 50 and 10 ppm for tricyclazole and Sudan I, respectively. The quaternized CH coating was essential to avoid AgNPs aggregation and create hot spots between interconnected AgNPs to provide a significant signal magnification [60].

Another interesting application of CH based systems addresses the growing awareness towards water purification. Ramalingam et al. [61] developed core-shell NPs with a Fe_3O_4 magnetic core and a CH shell which were further decorated with AgNPs, for adsorptive removal of dyes and microbial contaminants from water (Fig. 2.2). In this multifunctional hybrid system, the magnetic core allowed the easy separation of the material using an external magnetic field leading to the recycling and reuse, while CH favoured the binding of dyes, and the AgNPs inhibited the bacterial growth and prevented biofilm formation on the microspheres. This superparamagnetic hybrid material removed efficiently the microbial contaminants and at the same time 99.5% of dyes on single and multi-component systems showing a superior adsorption capacity when compared with other commons adsorbents [61].

A similar procedure was used by Xu et al. [63], however, in this case, the purpose was to prepare an efficient catalytic material. Herein, CH played the dual role of encapsulating the magnetic core and acting as reducing agent. The hybrid microcapsules were used as catalysts in the reduction of p-nitrophenol to p-aminophenol with $NaBH_4$, exhibiting a conversion efficiency of 98% within 15 min. Moreover, the catalysts can be recycled and reused successfully for at least ten cycles.

Hybrid materials based on CH and CuNPs have a demonstrated applicability as catalytic systems. As an illustrative example, Frindy et al. [64] prepared porous microspheres where the CuNPs were embedded within the CH matrix (Cu maximum loading of 2.3 wt%). These aerogels are effective heterogeneous catalysts for the C–S coupling of aryl halides and thiophenol in toluene. The catalyst was more active for aryl iodides than for aryl bromides and chlorides and can be reused up to four times, under optimal conditions. Nevertheless, this material showed some disadvantages, namely Cu leaching from the first to the second use, and the "poisoning activity" from the halides released during the reaction that increased the Cu leaching [64].

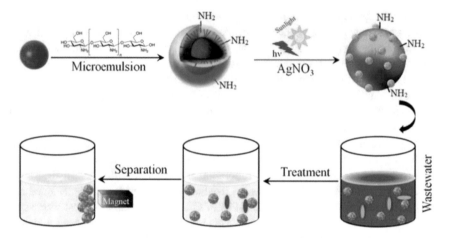

Fig. 2.2 Synthesis of AgNPs decorated core-shell type magnetic-CH microspheres for application in water purification. Reprinted with permission from [61]. Copyright 2015 American Chemical Society

The preparation of antifouling coatings against the growth of algae [65] and sorbent membranes for the removal of chromate and vanadate from aquatic environment [66] are other applications reported for the CH/CuNPs hybrid materials. Following a distinct perspective, Gómez et al. [67] developed an interesting CH/CuNPs hybrid material for utilization in modern agriculture, aiming to improve plants development and growth following a more ecological way through a grafting cultivation technique. CH/poly(vinyl alcohol) hydrogels soaked with CuNPs (dilute solutions) were grafted on watermelon plants ("Jubilee" cultivar) via the tongue approach method, in order to assess the changes in the grafted plant growth and stomatal morphology (density, index, width and length). The leaf micromorphology was changed with an increase in the growth of the primary stems, the root system, and in stomatal width. This positive behaviour was explained by the authors as a probable improvement in the ability to take up water and nutrients from their environment, which is translated into a superior growth rate [67].

As verified for most of palladium-based materials, the driving force for the development of CH/PdNPs materials is based on the effective catalytic properties of this metallic NPs. For this specific type of hybrid materials, the focus was the preparation of heterogeneous nanocatalysts for Suzuki–Miyaura coupling reactions. The importance of this carbon-carbon bond formation reaction lies, in the fact, that is one of the most important methods for the synthesis of biaryl products in a single-step process. Usually, CH was modified with distinct molecules *e.g.*, biguanidine [68] and thiourea [69], before being used to functionalize the PdNPs. The objective of the CH functionalization is to provide coordination sites for immobilization of PdNPs. Veisi et al. [68] functionalized CH with dicyandiamide to obtain biguanidine groups used to adsorb Pd (II) ions which were later reduced using hydrazine hydrate. This metal-polymer hybrid material performed as an excellent catalyst for Suzuki coupling reactions in terms of activity for various aryl halides, including less reactive chlorobenzenes. The hybrid reusability was also demonstrated with a high catalytic activity even after six runs and without metal leaching [68]. Affrose et al. [69] followed a dissimilar procedure to prepare CH/PdNPs hybrid materials using an aqueous system and a green reducing agent, *viz.* ellagic acid; hence, avoiding organic solvents and the use of hydrazine. The reaction was performed with various heterocyclic boronic acids and the catalyst could also be easily recovered and reused for at least five runs without losing its activity.

In a different vein, Zeng et al. [70] prepared a versatile platform of CH and montmorillonite, that works as a suitable scaffolding material to incorporate PdNPs. This porous matrix allowed the easy diffusion of reactants and respective product molecules, showing to be highly active for the Heck reactions of aromatic halides and alkenes. This economical and abundant heterogeneous catalyst demonstrated to be recycled thirty times without significant loss of activity [70].

In the biomedical field, Bharathiraja et al. [71] give a step forward using PdNPs modified with thiolated CH (confers biocompatibility and enables further functionalization with other molecules via conventional coupling chemistry using the amine and hydroxyl groups present in the polymer) to develop a nanotheranostic agent for enhanced imaging and therapy of tumours tissues using a near-infrared laser (Fig. 2.3). The PdNPs were functionalised with CH to improve the biocompatibility of the particles and an arginine-glycine-aspartic acid (RGD) peptide to increase the mNPs accumulation in the cancer cells and thus enhancing the respective photothermal therapeutic effects. The resulting hybrid material showed good biocompatibility, water dispersity, colloidal and physiological stability, and capability to destroy the tumour effectively under 808 nm laser illumination (photothermal transduction efficiency was comparable with the presented by isolated Au nanorods, as Au is the standard reference mNPs). Furthermore, the material gives a good amplitude of photoacoustic signals, enabling the tumour tissues imaging using a non-invasive photoacoustic tomography system. However, the authors alert to the significant challenges that remain in advancing from laboratory settings to clinical therapy. In this case, the biggest challenge will be the need to ensure the reproducibility and uniformity of these functionalized PdNPs after the scale-up of the process [71].

2.4 Starch/mNPs Hybrid Materials

Starch is another polysaccharide that can be used to develop hybrid materials with mNPs, as summarized in Table 2.3 showing recent examples of the combination between starch and different mNPs, namely AuNPs [72–76], AgNPs [77–84], CuNPs [85, 86] and PdNPs [87–91].

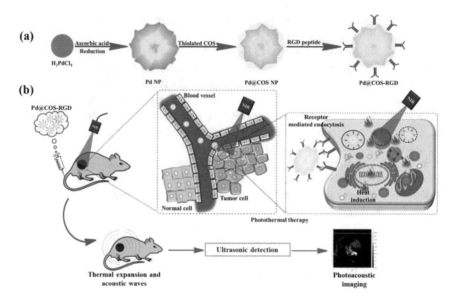

Fig. 2.3 a Preparation of PdNPs and further surface coating with thiolated CH and, functionalization using RGD peptide. **b** Photothermal ablation and photoacoustic imaging of tumour tissue using CH/PdNPs hybrid material. Reprinted with permission from [71]. Copyright 2018 Springer Nature

Usually, the methodology used for the preparation of starch/AuNPs hybrid systems comprises the reduction of an Au salt by distinct reduction agents (*e.g.* sodium citrate) in the presence of starch that acts as stabilizer. The hybrids prepared following this methodology envisaged, for example, the development of colourimetric sensing materials that can be used for the detention of heavy metals, such as Cu^{2+} and Pb^{2+} in contaminated waters [72] or for the detection of the protein content in milk [73]. Another application described for starch-stabilized gold materials includes the development of catalytic systems, namely for the homocoupling of phenylboronic acid in water using oxygen in air as oxidant at ambient temperature [74]. The same authors developed later a similar catalytic material for the degradation of 4-nitrophenol, however, in this case, a green synthesis method based on the use of mung bean starch as reducing (aldehyde functional terminal groups) and stabilizing (hydroxyl groups) agent was used to prepare the AuNPs hybrid material [75].

A distinct methodology was described by Pagno et al. [76] for the preparation of hybrid biofilms from quinoa starch with potential application as active food packaging materials. In this work, cast biofilms with different gold NPs contents where obtained through the mixing of AuNPs, stabilised by an ionic silsesquioxane, with a 4% starch suspension, using glycerol as plasticizer. These hybrid biofilms demonstrated an increased tensile strength, UV radiation absorption, thermal stability, antimicrobial activity and a decreased solubility in relation to the pure starch biofilm.

Table 2.3 Examples of hybrid materials based on starch and mNPs, preparation methodologies and potential applications

Metal NPs	Methodology	Application	References
Au	Reduction of Au salt with NaOH in presence of starch	Sensors (heavy metals detection)	[72]
	Reduction of Au salt with NaOH in presence of starch	Sensors (colorimetric detector)	[73]
	Reduction of Au^{3+} with a starch complex with sodium citrate	Catalysis (homocoupling of phenylboronic acid)	[74]
	Reduction of Au salt using starch	Catalysis (reduction of 4-nitrophenol)	[75]
	Biofilms prepared by casting	Food packaging	[76]
Ag	Reduction of Ag salt with dextrose in presence of starch	Sensors (colorimetric detection of hydrogen peroxide)	[77]
	Reduction of Ag salt with $NaBH_4$ in presence of starch	Biomedical (tissue regeneration applications)	[78]
	Biofilms prepared by microwave-assisted syntheses	Antibacterial materials	[79]
	Reduction of Ag salt with sodium citrate in presence of starch	Sensors (analysis of food and environmental samples)	[80]
	Reduction of Ag salt using starch under sonication	Catalysis (synthesis of 2-aryl substituted benzimidazoles)	[81]
	Blending of the components using glycerol as plasticizer	Antimicrobial packaging, biomedicine and sensors	[82, 83]
	Blending of the components using glycerol as plasticizer	Antimicrobial packaging	[84]
Cu	Deposition of CuNPs on starch microparticles	Catalysis	[85]
	Blending of the components	Antimicrobial hydrogels	[86]
Pd	Reduction of Pd salt with sodium borohydride in presence of starch/chitosan	Catalysis (synthesis of biphenyl compounds via Suzuki-Miyaura reactions)	[87]
	Formation of PdNPs in presences of starch and NaOH	Catalysis (Suzuki-Miyaura cross-coupling reactions)	[88]
	Reduction of Pd salt with citric acid in the presence of starch	Catalysis (Suzuki and Heck cross-coupling reactions)	[89]
	Grafting of starch on the surface of PdNPs	Catalysis (Heck and Sonogashira coupling reactions)	[90]
	Mixture of PdNPs with amino-functionalized starch	Catalysis (oxidation of alcohols)	[91]

As verified for other polysaccharides, AgNPs-based starch hybrids are also the most common materials reported in literature. Most of these works combine starch with distinct reducing agents for the preparation of starch-capped AgNPs hybrid systems, where starch acts as a stabilizing, shape-directing and/or capping agent during the growth process of the nanostructures. One example is the work reported by Mohan et al. [77] where starch and dextrose were used as stabilizing and reducing agent, respectively. The as-prepared NPs showed antibacterial activity but, at same time, the ability to the colourimetric detection of hydrogen peroxide (down to concentrations of 1×10^{-10} M). A similar work was carried out by Mandal and co-workers who prepared stable AgNPs capped with variable concentrations of sago starch [78]. These nanostructures were incorporated into collagen extracted from fish scales and freeze-dried to form scaffolds for tissue engineering. The scaffolds display enhanced stability and improved Young's modulus when compared with pristine collagen scaffolds. The in vitro studies showed that these materials are biocompatible and presented a strong antibacterial activity.

Kahrilas et al. [79] prepared AgNPs caped with starch by a microwave-assisted methodology. Following this green approach, NPs with a size of 12.1 ± 4.8 nm were obtained in less than 15 min exhibiting a clear antibacterial effect on a variety of Gram-positive and Gram-negative bacteria (*E. coli, Bacillus subtilis, Klebsiella pneumoniae, P. aeruginosa, S. aureus*, and *Janthinobacterium lividum*). These hybrid systems could be applied in the development of distinct antimicrobial materials. The authors pointed the further steps in this field, namely the evaluation of the effects of AgNPs versus antibiotics, and the effect of AgNPs on chemical signalling of specific compounds or metabolites. Other interesting applications for starch coated AgNPs involve their use in sensors for the analysis of food and environmental samples using SERS [80] or in catalysis for the synthesis of 2-aryl substituted benzimidazoles, which are compounds with numerous biomedical applications [81].

Starch/AgNPs hybrid materials can also be processed in the form of films for application as packaging materials. The work reported by Cheviron et al. [82, 83] showed this possibility, where the as-prepared AgNPs were dispersed in a potato starch/glycerol matrix and then casted. However, regardless of the silver incorporation approach, no significant difference on the thermal stability of hybrid films was observed. However, a lower water-uptake and a high decrease of relative water and oxygen permeability were detected when compared to the associated neat matrix. More recently, and following a similar procedure, Ortega et al. [84] prepared hybrid films, but in addition to the study of the abovementioned properties, the mechanical properties, the antimicrobial activity, the heat-sealing capacity, and the conservation of a dairy product (fresh cheese) protected by the hybrid films were also evaluated. As verified in the previous studies [82, 83], a decrease in water vapour permeability with increasing concentration of AgNPs was observed. Besides, AgNPs incorporation contributed to the matrix reinforcement, strong antimicrobial activity, and to extend the shelf-life of fresh cheese samples by 21 days. This example demonstrates that the starch active films have a high potential to be used as active food packaging materials, however, as referred by the authors, tests on the AgNPs toxicity and migration to the product are still necessary [84].

For starch/PdNPs based hybrid materials, the most common application is once again on catalysis due to the well-known catalytic activity of palladium. The described works showed the applicability of these systems in Suzuki-Miyaura [87–89] and Heck and Sonogashira [89, 90] cross-coupling reactions, and selective oxidation of primary alcohols to aldehydes [91]. Usually, the hybrid materials are prepared either by synthesizing the PdNPs in the presence of starch and using $NaBH_4$ or citric acid as reducing agents, or by synthesizing the PdNPs with the previous reducing agents and then graft the polysaccharide moieties into the surface of the NPs.

A recent illustrative example was described by Tukhani et al. [90], that developed a catalytic system through the immobilization of PdNPs on starch-functionalized magnetic NPs (Fe_3O_4 NPs) coated with a silica layer and functionalized with chlorosilyl groups at their surface (Fig. 2.4). This magnetic and reusable catalystic system obtained by a "green" approach is active in Heck and Sonogashira coupling reactions in water, demonstrating a good reusability with no significant loss of catalytic activity after 5 cycles.

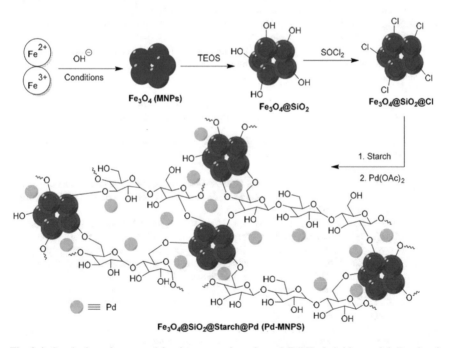

Fig. 2.4 Synthetic pathway used for the preparation of starch/PdNPs hybrid material. Reprinted with permission from [90]. Copyright 2018 American Chemical Society

References

1. Sciau P. Nanoparticles in ancient materials: the metallic lustre decorations of medieval ceramics. In: The delivery of nanoparticles. InTech; 2012. p. 525–40.
2. Liz-Marzán LM. Tailoring surface plasmons through the morphology and assembly of metal nanoparticles. Langmuir. 2006;22:32–41.
3. Caseri W. Nanocomposites of polymers and metals or semiconductors: historical background and optical properties. Macromol Rapid Commun. 2000;21:705–22.
4. You H, Yang S, Ding B, Yang H. Synthesis of colloidal metal and metal alloy nanoparticles for electrochemical energy applications. Chem Soc Rev. 2013;42:2880–904.
5. Bhattacharya R, Mukherjee P. Biological properties of "naked" metal nanoparticles. Adv Drug Deliv Rev. 2008;60:1289–306.
6. Islam MS, Chen L, Sisler J, Tam KC. Cellulose nanocrystal (CNC)–inorganic hybrid systems: synthesis, properties and applications. J Mater Chem B. 2018;6:864–83.
7. Pinto RJB, Neves MC, Neto CP, Trindade T. Composites of cellulose and metal nanoparticles. In: Nanocomposites—new trends and developments. InTech; 2012. p. 73–96.
8. Foresti ML, Vázquez A, Boury B. Applications of bacterial cellulose as precursor of carbon and composites with metal oxide, metal sulfide and metal nanoparticles: a review of recent advances. Carbohydr Polym. 2017;157:447–67.
9. Kaushik M, Moores A. Review: nanocelluloses as versatile supports for metal nanoparticles and their applications in catalysis. Green Chem. 2016;18:622–37.
10. Ge S, Zhang L, Zhang Y, Lan F, Yan M, Yu J. Nanomaterials-modified cellulose paper as a platform for biosensing applications. Nanoscale. 2017;9:4366–82.
11. Rajwade JM, Paknikar KM, Kumbhar JV. Applications of bacterial cellulose and its composites in biomedicine. Appl Microbiol Biotechnol. 2015;99:2491–511.
12. Van Rie J, Thielemans W. Cellulose–gold nanoparticle hybrid materials. Nanoscale. 2017;9:8525–54.
13. Chen Y, Chen S, Wang B, Yao J, Wang H. TEMPO-oxidized bacterial cellulose nanofibers-supported gold nanoparticles with superior catalytic properties. Carbohydr Polym. 2017;160:34–42.
14. Wu X, Lu C, Zhou Z, Yuan G, Xiong R, Zhang X. Green synthesis and formation mechanism of cellulose nanocrystal-supported gold nanoparticles with enhanced catalytic performance. Environ Sci Nano. 2014;1:71–9.
15. Li Y, Tseng Y, Unnikrishnan B, Huang C. Gold-Nanoparticles-modified cellulose membrane coupled with laser desorption/ionization mass spectrometry for detection of iodide in urine. ACS Appl Mater Interfaces. 2013;5:9161–6.
16. Almeida A, Rosa AMM, Azevedo AM, Prazeres DMF. A biomolecular recognition approach for the functionalization of cellulose with gold nanoparticles. J Mol Recognit. 2017;30:e2634.
17. Chu G, Wang X, Yin H, Shi Y, Jiang H, Chen T, Gao J, Qu D, Xu Y, Ding D. Free-standing optically switchable chiral plasmonic photonic crystal based on self-assembled cellulose nanorods and gold nanoparticles. ACS Appl Mater Interfaces. 2015;7:21797–806.
18. Majoinen J, Hassinen J, Haataja JS, Rekola HT, Kontturi E, Kostiainen MA, Ras RHA, Törmä P, Ikkala O. Chiral plasmonics using twisting along cellulose nanocrystals as a template for gold nanoparticles. Adv Mater. 2016;28:5262–7.
19. Basavaraja C, Kim JK, Huh DS. Characterization and temperature-dependent conductivity of polyaniline nanocomposites encapsulating gold nanoparticles on the surface of carboxymethyl cellulose. Mater Sci Eng, B. 2013;178:167–73.
20. Faria-Tischer PCS, Costa CAR, Tozetti I, Dall'Antonia LH, Vidotti M. Structure and effects of gold nanoparticles in bacterial cellulose–polyaniline conductive membranes. RSC Adv. 2016;6:9571–80.
21. Bumbudsanpharoke N, Choi J, Park I, Ko S. Facile biosynthesis and antioxidant property of nanogold-cellulose fiber composite. J Nanomater. 2015;2015:1–9.

22. Li Y, Tian Y, Zheng W, Feng Y, Huang R, Shao J, Tang R, Wang P, Jia Y, Zhang J, Zheng W, Yang G, Jiang X. Composites of bacterial cellulose and small molecule-decorated gold nanoparticles for treating gram-negative bacteria-infected wounds. Small. 2017;13:1700130.

23. Wu J, Zheng Y, Song W, Luan J, Wen X, Wu Z, Chen X, Wang Q, Guo S. In situ synthesis of silver-nanoparticles/bacterial cellulose composites for slow-released antimicrobial wound dressing. Carbohydr Polym. 2014;102:762–71.

24. Singla R, Soni S, Kulurkar PM, Kumari A, Mahesh S, Patial V, Padwad YS, Yadav SK. In situ functionalized nanobiocomposites dressings of bamboo cellulose nanocrystals and silver nanoparticles for accelerated wound healing. Carbohydr Polym. 2017;155:152–62.

25. Wang K, Ma Q, Wang SD, Liu H, Zhang SZ, Bao W, Zhang KQ, Ling LZ. Electrospinning of silver nanoparticles loaded highly porous cellulose acetate nanofibrous membrane for treatment of dye wastewater. Appl Phys A Mater Sci Process. 2016;122:1–10.

26. Praveena SM, Karuppiah K, Than LTL. Potential of cellulose paper coated with silver nanoparticles: a benign option for emergency drinking water filter. Cellulose. 2018;25:2647–58.

27. Tang J, Shi Z, Berry RM, Tam KC. Mussel-inspired green metallization of silver nanoparticles on cellulose nanocrystals and their enhanced catalytic reduction of 4-nitrophenol in the presence of β-cyclodextrin. Ind Eng Chem Res. 2015;54:3299–308.

28. Liou P, Nayigiziki FX, Kong F, Mustapha A, Lin M. Cellulose nanofibers coated with silver nanoparticles as a SERS platform for detection of pesticides in apples. Carbohydr Polym. 2017;157:643–50.

29. Zhou S, Wang M, Chen X, Xu F. Facile template synthesis of microfibrillated cellulose/polypyrrole/silver nanoparticles hybrid aerogels with electrical conductive and pressure responsive properties. ACS Sustain Chem Eng. 2015;3:3346–54.

30. Solomon MM, Gerengi H, Umoren SA. Carboxymethyl cellulose/silver nanoparticles composite: synthesis, characterization and application as a benign corrosion inhibitor for St37 steel in 15% H_2SO_4 medium. ACS Appl Mater Interfaces. 2017;9:6376–89.

31. Zhong T, Oporto GS, Jaczynski J, Jiang C. Nanofibrillated cellulose and copper nanoparticles embedded in polyvinyl alcohol films for antimicrobial applications. Biomed Res Int. 2015;2015:1–8.

32. Sadanand V, Rajini N, Varada Rajulu A, Satyanarayana B. Preparation of cellulose composites with in situ generated copper nanoparticles using leaf extract and their properties. Carbohydr Polym. 2016;150:32–9.

33. Al-Enizi AM, Ahamad T, Al-hajji AB, Ahmed J, Chaudhary AA, Alshehri SM. Cellulose gum and copper nanoparticles based hydrogel as antimicrobial agents against urinary tract infection (UTI) pathogens. Int J Biol Macromol. 2018;109:803–9.

34. Bajpai S, Das P, Soni B. Copper nanoparticles loaded cellulose-g-poly acrylic acid fibers with antibacterial properties. J Ind Text. 2016;45:495–515.

35. Lv P, Wei A, Wang Y, Li D, Zhang J, Lucia LA, Wei Q. Copper nanoparticles-sputtered bacterial cellulose nanocomposites displaying enhanced electromagnetic shielding, thermal, conduction, and mechanical properties. Cellulose. 2016;23:3117–27.

36. Baruah D, Pahari P, Konwar D. Synthesis of (E)-nitroolefins and substituted nitrobenzenes via decarboxylative nitration using cellulose supported copper nanoparticles. Tetrahedron Lett. 2015;56:2418–21.

37. Rezayat M, Blundell RK, Camp JE, Walsh DA, Thielemans W. Green one-step synthesis of catalytically active palladium nanoparticles supported on cellulose nanocrystals. ACS Sustain Chem Eng. 2014;2:1241–50.

38. Li G, Li Y, Wang Z, Liu H. Green synthesis of palladium nanoparticles with carboxymethyl cellulose for degradation of azo-dyes. Mater Chem Phys. 2017;187:133–40.

39. Ahmar H, Keshipour S, Hosseini H, Fakhari AR, Shaabani A, Bagheri A. Electrocatalytic oxidation of hydrazine at glassy carbon electrode modified with ethylenediamine cellulose immobilized palladium nanoparticles. J Electroanal Chem. 2013;690:96–103.

40. Sultana T, Mandal BH, Rahman ML, Sarkar SM. Bio-waste corn-cob cellulose supported poly(amidoxime) palladium nanoparticles for suzuki-miyaura cross-coupling reactions. Chem SELECT. 2016;1:4108–12.

41. Huang Y, Fang Y, Chen L, Lu A, Zhang L. One-step synthesis of size-tunable gold nanoparticles immobilized on chitin nanofibrils via green pathway and their potential applications. Chem Eng J. 2017;315:573–82.
42. Singh R, Singh D. Chitin membranes containing silver nanoparticles for wound dressing application. Int Wound J. 2014;11:264–8.
43. Nguyen VQ, Ishihara M, Mori Y, Nakamura S, Kishimoto S, Hattori H, Fujita M, Kanatani Y, Ono T, Miyahira Y, Matsui T. Preparation of size-controlled silver nanoparticles and chitin-based composites and their antimicrobial activities. J Nanomater. 2013;2013:1–7.
44. Ifuku S, Tsukiyama J, Yukawa T, Egusa M, Kaminaka H, Izawa H, Morimoto M, Saimoto H. Facile preparation of silver nanoparticles immobilized on chitin nanofiber surfaces to endow antifungal activities. Carbohydr Polym. 2015;117:813–7.
45. El Kadib A, Bousmina M, Brunel D. Recent progress in chitosan bio-based soft nanomaterials. J Nanosci Nanotechnol. 2014;14:308–31.
46. Dutta PK, Srivastava R, Dutta J. Functionalized nanoparticles and chitosan-based functional nanomaterials. In: Multifaceted development and application of biopolymers for biology, biomedicine and nanotechnology. 2012. p. 349–59.
47. Ahmed T, Aljaeid B. Preparation, characterization, and potential application of chitosan, chitosan derivatives, and chitosan metal nanoparticles in pharmaceutical drug delivery. Drug Des Dev Ther. 2016;10:483.
48. Bui V, Park D, Lee Y-C. Chitosan combined with ZnO, TiO$_2$ and Ag nanoparticles for antimicrobial wound healing applications: a mini review of the research trends. Polymers. 2017;9:21.
49. Lee M, Chen B-Y, Den W. Chitosan as a natural polymer for heterogeneous catalysts support: a short review on its applications. Appl Sci. 2015;5:1272–83.
50. Dervisevic M, Dervisevic E, Çevik E, Şenel M. Novel electrochemical xanthine biosensor based on chitosan–polypyrrole–gold nanoparticles hybrid bio-nanocomposite platform. J Food Drug Anal. 2017;25:510–9.
51. Güner A, Çevik E, Şenel M, Alpsoy L. An electrochemical immunosensor for sensitive detection of Escherichia coli O157:H7 by using chitosan, MWCNT, polypyrrole with gold nanoparticles hybrid sensing platform. Food Chem. 2017;229:358–65.
52. Manivasagan P, Bharathiraja S, Bui NQ, Lim IG, Oh J. Paclitaxel-loaded chitosan oligosaccharide-stabilized gold nanoparticles as novel agents for drug delivery and photoacoustic imaging of cancer cells. Int J Pharm. 2016;511:367–79.
53. Fathi M, Sahandi Zangabad P, Barar J, Aghanejad A, Erfan-Niya H, Omidi Y. Thermo-sensitive chitosan copolymer-gold hybrid nanoparticles as a nanocarrier for delivery of erlotinib. Int J Biol Macromol. 2018;106:266–76.
54. Ahmed RA, Fadl-Allah SA, El-Bagoury N, El-Rab SMFG. Improvement of corrosion resistance and antibacterial effect of NiTi orthopedic materials by chitosan and gold nanoparticles. Appl Surf Sci. 2014;292:390–9.
55. Tentor FR, de Oliveira JH, Scariot DB, Lazarin-Bidóia D, Bonafé EG, Nakamura CV, Venter SAS, Monteiro JP, Muniz EC, Martins AF. Scaffolds based on chitosan/pectin thermosensitive hydrogels containing gold nanoparticles. Int J Biol Macromol. 2017;102:1186–94.
56. Kostevsek N, Locatelli E, Garrovo C, Arena F, Monaco I, Nikolov IP, Sturm S, Zuzek Rozman K, Lorusso V, Giustetto P, Bardini P, Biffi S, Comes Franchini M. The one-step synthesis and surface functionalization of dumbbell-like gold–iron oxide nanoparticles: a chitosan-based nanotheranostic system. Chem Commun. 2016;52:378–81.
57. Holubnycha V, Kalinkevich O, Ivashchenko O, Pogorielov M. Antibacterial activity of in situ prepared chitosan/silver nanoparticles solution against methicillin-resistant strains of Staphylococcus aureus. Nanoscale Res Lett. 2018;13:71.
58. Shao J, Yu N, Kolwijck E, Wang B, Tan KW, Jansen JA, Walboomers XF, Yang F. Biological evaluation of silver nanoparticles incorporated into chitosan-based membranes. Nanomedicine. 2017;12:2771–85.
59. Potara M, Boca S, Licarete E, Damert A, Alupei M-C, Chiriac MT, Popescu O, Schmidt U, Astilean S. Chitosan-coated triangular silver nanoparticles as a novel class of biocompatible, highly sensitive plasmonic platforms for intracellular SERS sensing and imaging. Nanoscale. 2013;5:6013–22.

60. Chen K, Shen Z, Luo J, Wang X, Sun R. Quaternized chitosan/silver nanoparticles composite as a SERS substrate for detecting tricyclazole and Sudan I. Appl Surf Sci. 2015;351:466–73.
61. Ramalingam B, Khan MMR, Mondal B, Mandal AB, Das SK. Facile synthesis of silver nanoparticles decorated magnetic-chitosan microsphere for efficient removal of dyes and microbial contaminants. ACS Sustain Chem Eng. 2015;3:2291–302.
62. Djerahov L, Vasileva P, Karadjova I, Kurakalva RM, Aradhi KK. Chitosan film loaded with silver nanoparticles—sorbent for solid phase extraction of Al(III), Cd(II), Cu(II), Co(II), Fe(III), Ni(II), Pb(II) and Zn(II). Carbohydr Polym. 2016;147:45–52.
63. Xu P, Liang X, Chen N, Tang J, Shao W, Gao Q, Teng Z. Magnetic separable chitosan microcapsules decorated with silver nanoparticles for catalytic reduction of 4-nitrophenol. J Colloid Interface Sci. 2017;507:353–9.
64. Frindy S, El Kadib A, Lahcini M, Primo A, García H. Copper nanoparticles stabilized in a porous chitosan aerogel as a heterogeneous catalyst for C-S cross-coupling. ChemCatChem. 2015;7:3307–15.
65. Abiraman T, Balasubramanian S. Synthesis and characterization of large-scale (<2 nm) chitosan-decorated copper nanoparticles and their application in antifouling coating. Ind Eng Chem Res. 2017;56:1498–508.
66. de Godoi FC, Rodriguez-Castellon E, Guibal E, Beppu MM. An XPS study of chromate and vanadate sorption mechanism by chitosan membrane containing copper nanoparticles. Chem Eng J. 2013;234:423–9.
67. Gómez HG, Godina FR, Ortiz HO, Mendoza AB, Torres VR, De la Fuente MC. Use of chitosan-pva hydrogels with copper nanoparticles to improve the growth of grafted watermelon. Molecules. 2017;22:1031.
68. Veisi H, Ghadermazi M, Naderi A. Biguanidine-functionalized chitosan to immobilize palladium nanoparticles as a novel, efficient and recyclable heterogeneous nanocatalyst for Suzuki-Miyaura coupling reactions. Appl Organomet Chem. 2016;30:341–5.
69. Affrose A, Suresh P, Azath IA, Pitchumani K. Palladium nanoparticles embedded on thiourea-modified chitosan: a green and sustainable heterogeneous catalyst for the Suzuki reaction in water. RSC Adv. 2015;5:27533–9.
70. Zeng M, Wang Y, Liu Q, Yuan X, Zuo S, Feng R, Yang J, Wang B, Qi C, Lin Y. Encaging palladium nanoparticles in chitosan modified montmorillonite for efficient, recyclable catalysts. ACS Appl Mater Interfaces. 2016;8:33157–64.
71. Bharathiraja S, Bui NQ, Manivasagan P, Moorthy MS, Mondal S, Seo H, Phuoc NT, Vy Phan TT, Kim H, Lee KD, Oh J. Multimodal tumor-homing chitosan oligosaccharide-coated biocompatible palladium nanoparticles for photo-based imaging and therapy. Sci Rep. 2018;8:500.
72. Ban DK, Pratihar SK, Paul S. Controlled modification of starch in the synthesis of gold nanoparticles with tunable optical properties and their application in heavy metal sensing. RSC Adv. 2015;5:81554–64.
73. Vantasin S, Pienpinijtham P, Wongravee K, Thammacharoen C, Ekgasit S. Naked eye colorimetric quantification of protein content in milk using starch-stabilized gold nanoparticles. Sens Actuators B Chem. 2013;177:131–7.
74. Wongmanee K, Khuanamkam S, Chairam S. Gold nanoparticles stabilized by starch polymer and their use as catalyst in homocoupling of phenylboronic acid. J King Saud Univ Sci. 2017;29:547–52.
75. Chairam S, Konkamdee W, Parakhun R. Starch-supported gold nanoparticles and their use in 4-nitrophenol reduction. J Saudi Chem Soc. 2017;21:656–63.
76. Pagno CH, Costa TMH, de Menezes EW, Benvenutti EV, Hertz PF, Matte CR, Tosati JV, Monteiro AR, Rios AO, Flôres SH. Development of active biofilms of quinoa (*Chenopodium quinoa* W.) starch containing gold nanoparticles and evaluation of antimicrobial activity. Food Chem. 2015;173:755–62.
77. Mohan S, Oluwafemi OS, Songca SP, Jayachandran VP, Rouxel D, Joubert O, Kalarikkal N, Thomas S. Synthesis, antibacterial, cytotoxicity and sensing properties of starch-capped silver nanoparticles. J Mol Liq. 2016;213:75–81.

78. Mandal A, Sekar S, Seeni Meera KM, Mukherjee A, Sastry TP, Mandal AB. Fabrication of collagen scaffolds impregnated with sago starch capped silver nanoparticles suitable for biomedical applications and their physicochemical studies. Phys Chem Chem Phys. 2014;16:20175–83.
79. Kahrilas GA, Haggren W, Read RL, Wally LM, Fredrick SJ, Hiskey M, Prieto AL, Owens JE. Investigation of antibacterial activity by silver nanoparticles prepared by microwave-assisted green syntheses with soluble starch, dextrose, and arabinose. ACS Sustain Chem Eng. 2014;2:590–8.
80. Zhao Y, Tian Y, Ma P, Yu A, Zhang H, Chen Y. Determination of melamine and malachite green by surface-enhanced Raman scattering spectroscopy using starch-coated silver nanoparticles as substrates. Anal Methods. 2015;7:8116–22.
81. Kumar B, Smita K, Cumbal L, Debut A, Pathak RN. Sonochemical synthesis of silver nanoparticles using starch: a comparison. Bioinorg Chem Appl. 2014;2014:1–8.
82. Cheviron P, Gouanvé F, Espuche E. Green synthesis of colloid silver nanoparticles and resulting biodegradable starch/silver nanocomposites. Carbohydr Polym. 2014;108:291–8.
83. Cheviron P, Gouanvé F, Espuche E. Effect of silver nanoparticles' generation routes on the morphology, oxygen, and water transport properties of starch nanocomposite films. J Nanopart Res. 2015;17:364.
84. Ortega F, Giannuzzi L, Arce VB, García MA. Active composite starch films containing green synthetized silver nanoparticles. Food Hydrocolloids. 2017;70:152–62.
85. Gholinejad M, Saadati F, Shaybanizadeh S, Pullithadathil B. Copper nanoparticles supported on starch micro particles as a degradable heterogeneous catalyst for three-component coupling synthesis of propargylamines. RSC Adv. 2016;6:4983–91.
86. Villanueva ME, Diez AMDR, González JA, Pérez CJ, Orrego M, Piehl L, Teves S, Copello GJ. Antimicrobial activity of starch hydrogel incorporated with copper nanoparticles. ACS Appl Mater Interfaces. 2016;8:16280–8.
87. Baran T, Yılmaz Baran N, Menteş A. Sustainable chitosan/starch composite material for stabilization of palladium nanoparticles: synthesis, characterization and investigation of catalytic behaviour of Pd@chitosan/starch nanocomposite in Suzuki-Miyaura reaction. Appl Organomet Chem. 2018;32:e4075.
88. Dewan A, Bharali P, Bora U, Thakur AJ. Starch assisted palladium(0) nanoparticles as in situ generated catalysts for room temperature Suzuki-Miyaura reactions in water. RSC Adv. 2016;6:11758–62.
89. Patil AB, Bhanage BM. Solar energy assisted starch-stabilized palladium nanoparticles and their application in C—C coupling reactions. J Nanosci Nanotechnol. 2013;13:5061–8.
90. Tukhani M, Panahi F, Khalafi-Nezhad A. Supported palladium on magnetic nanoparticles–starch substrate (Pd-MNPSS): highly efficient magnetic reusable catalyst for C-C coupling reactions in water. ACS Sustain Chem Eng. 2018;6:1456–67.
91. Verma S, Tripathi D, Gupta P, Singh R, Bahuguna GM, Shivakumar KLN, Chauhan RK, Saran S, Jain SL. Highly dispersed palladium nanoparticles grafted onto nanocrystalline starch for the oxidation of alcohols using molecular oxygen as an oxidant. Dalton Trans. 2013;42:11522.

Chapter 3
Polysaccharides-Based Hybrids with Metal Oxide Nanoparticles

Metal oxide nanoparticles (NPs), such as titanium dioxide (TiO_2), zinc oxide (ZnO), copper oxides (CuO and Cu_2O), silica (SiO_2) and iron oxides (Fe_2O_3 and Fe_3O_4) NPs, are high in request for several high-tech applications, including catalysts, absorbent materials, optoelectronic materials, magnetic composites, luminescent materials, drug delivery systems, sensors, antimicrobial materials, imaging, among many others [1–5]. In this context, the association of metal oxide NPs with different materials [6–8] is a well-recognized approach to produce metal oxide-based hybrid nanomaterials with a wide range of properties and controlled morphologies with improved stability and better performance.

In this perspective, the partnership between polysaccharides and metal oxide NPs has been extensively explored in different fields of research [8], namely (i) on the conversion of polysaccharides into renewable based fuels and chemicals, (ii) on the transformation of polysaccharides into carbons, (iii) as modifiers of the metal oxides' surfaces, and (iv) on the development of novel hybrid nanomaterials. Boury and Plumejeau [8] reviewed the association of metal oxides and polysaccharides, with emphasis on the different methods to combine them, on the controlled hydrolysis/polycondensation and nucleation growth process of metal oxides NPs using polysaccharide fibres and templating (moulding raw materials with metal oxides) of polysaccharide fibres. Salama [9] summarized and discussed the efforts of using polysaccharides to produce hybrid materials with silica, and Chauhan et al. [10] reviewed the methodologies for the incorporation of metal oxides in paper matrices. However, so far there we could not find in the literature an overarching appraisal of the production and applications of polysaccharide/metal oxides NPs hybrid materials. Thus, the focus of this chapter will be on significant recent publications regarding the production and applications of innovative hybrid materials based on metal oxides (particularly TiO_2, ZnO, CuO, Cu_2O, SiO_2, Fe_3O_4 and Fe_2O_3) NPs and most abundant polysaccharides, viz. cellulose (Table 3.1), chitin and chitosan (Table 3.2), and starch (Table 3.3).

Table 3.1 Examples of cellulose-based hybrids with metal oxides NPs, the preparation methodologies and potential applications

Metal oxide	Cellulose substrate	Methodology	Application	References
TiO_2	Cotton fabrics	Dip-padding-curing	Self-cleaning textiles	[11]
	Cellulose fibres	In situ formation of TiO_2	Nanosorbents for Pb^{2+}	[12]
	CNCs	Blending	Active food packaging	[13]
	BC	Impregnation of N-F doped BC membrane with the TiO_2 suspension	Photocatalytic water disinfection against bacteria	[14]
	BC	Self-assembly	Production of H_2 by photocatalysis	[15]
	BC	Impregnation of the BC membrane with the TiO_2 suspension	Photocatalytic dye degradation	[16]
	Electrospun cellulose fibres	Impregnation of cellulose nanofibres with the TiO_2 sol	Ammonia gas sensor	[17]
	Pulp fibres	Chemical adhesion and physical adhesion	Photocatalytic decomposition o VOCs	[18]
	CNCs	Sol-gel process	UV filters for skin care products	[19]
	BC	Impregnation of BC nanofibres with the TiO_2 sol	Wound healing composites	[20]
	CA	Dispersion of TiO_2 nanotubes in CA solution, casting	Clinical applications	[21]
ZnO	CNFs	Electrostatic assembly	Paper coating formulations	[22]
	Pulp fibres	Blending of ZnO NPs with a cellulose suspension	Antibody immobilization materials	[23]
	BC	In situ formation of ZnO NPs	Wound healing and water disinfection	[24]
	Cellulose paper	Dispersion of a ZnO film over a paper surface	Oxygen sensor	[25]
	Cotton pulp	Blending of ZnO NPs with a cellulose solution	Glucose biosensor	[26]

(continued)

Table 3.1 (continued)

Metal oxide	Cellulose substrate	Methodology	Application	References
	Cotton linter pulp	One-step coagulation process	Functional biomaterials	[27]
	CNFs	Blending of commercial ZnO NPs with cellulose suspensions	Wood coatings	[28]
	CNCs	Blending of ZnO NPs with a polymerizable solution	Water purification	[29]
	CNFs	In situ formation of ZnO NPs followed by self-assembly of hybrid nanofibres	Active food packaging	[30]
	BC	Impregnation of BC with a ZnO NPs suspension	Wound healing membranes	[31]
	CNCs	Copolymerization ice-template method	Water purification	[32]
	α-Cellulose	Dissolution and regeneration of cellulose followed by hydrothermal synthesis of ZnO	Photocatalysis	[33]
	Cotton fabrics	Blending of ZnO NPs (and other metal oxides) with finishing formulations	Functional textiles	[34]
	CA	Blending of ZnO NPs with a cellulose acetate solution	Photocatalysis, degradation of dyes	[35]
SiO_2	Cellulose microfibres	In situ sol-gel synthesis	Sensors for H_2O_2 detection	[36]
	Microcrystalline cellulose	Dispersion of SiO_2 NPs in a cellulose solution	Packaging materials	[37]
	CNFS	Dispersion of SiO_2 NPs in a CNFs suspension followed by vacuum filtration	Separators for lithium batteries	[38]
	Filter paper	In situ sol-gel synthesis	Colorimetric bioassays	[39]
	Cotton linter pulp	Dispersion of SiO_2 NPs in a cellulose solution	Sensor for moisture detection	[40]

(continued)

Table 3.1 (continued)

Metal oxide	Cellulose substrate	Methodology	Application	References
	CNFs	In situ sol-gel synthesis	Insulation materials	[41]
	Microcrystalline cellulose	Conjugation of cellulose with SiO_2 NPs by esterification	Nanotherapeutics for cancer treatment	[42]
	BC	In situ sol-gel synthesis of SiO_2 NPs	Various applications	[43]
	CA	Dispersion of SiO_2 NPs in a cellulose acetate mixture	Insulation material	[44]
	CNFs	Mixture of SiO_2 NPs with a cellulose suspension followed by wet spinning	Flame retardancy materials	[45]
	Microfibrillated cellulose	Mixture of SiO_2 NPs with a cellulose suspension followed	Filters, absorbers and catalysts	[46]
	CNCs	Directional freeze casting	Electrical insulators	[47]
	CA	Covalent immobilization of SiO_2 NPs in a cellulose membrane	Water purification	[48]
Iron oxides	CNCs	In situ co-precipitation method	Separation of proteins	[49]
	CNCs	Grafting of β-cyclodextrin onto the surface of $Fe_3O_4@SiO_2$ NPs	Removal of pharmaceutical residues	[50]
	CMC	LbL assembly of polyelectrolytes on F_3O_4 NPs surface	Drug delivery and protein immobilization	[51]
	CMC	In situ synthesis of Fe_3O_4 NPs	NMR contrast agents (imaging)	[52]
	Cellulose from native wheat straw	Template synthesis combined with co-precipitation	Removal of Cr(VI) from contaminated waters	[53]
	CMC	Conjugation of NPs with CMC by chemical reaction	Drug delivery and hyperthermia treatment	[54]

(continued)

Table 3.1 (continued)

Metal oxide	Cellulose substrate	Methodology	Application	References
	Methyl cellulose	In situ formation of magnetite NPs	Magnetic hyperthermia treatment (cancer treatment)	[55]
	Cotton linter pulp	Extrusion dropping technology from NaOH/urea aqueous solution	Removal of metal ions (water purification)	[56]
	CNCs	In situ synthesis of iron oxide	Flexible NO_2 sensors	[57]
	Hydroxyethyl cellulose	Fe_3O_4 NPs coating with HEC	Targeted drug delivery	[58]
	CNCs	Coupling reaction	Platforms for separation of proteins	[59]
	Filter paper	Deposition of magnetite NPs on cellulose paper sheets	Magnetic paper-based ELISA for IgM-dengue detection	[60]
Copper oxide CuO	CNCs	In situ formation of CuO NPs	Catalysis; reduction of 4-nitrophenol	[61]
	Cotton fabrics	In situ formation of CuO NPs	Wound dressing and other biomedical applications	[62]
	Cellulose paper	In situ formation of CuO NPs	Water purification	[63]
	Cotton fabrics	In situ formation of CuO NPs	Medical and technical textiles	[64]
	CMC	Synthesis of CuO followed by mixing with the polymer solutions and then casting	UV-light barrier and antibacterial food packaging	[65]
	CNCs	In situ formation of CuO NPs	Flexible supercapacitor electrodes	[66]
	BC	Adsorption of commercial CuO NPs	Water purification and food packaging	[67]

Table 3.2 Examples of chitosan-based hybrids with metal oxides NPs, the preparation methodologies and potential applications

Metal oxide	Type of material	Methodology	Application	References
TiO_2	Thin coating	Mixture of CH solution with TiO_2 NPs followed by coating of the substrates	Photocatalysis (degradation of the herbicide terbuthylazine)	[68]
	Core-shell microparticles	Molecular imprinting	Photocatalysis (degradation of methyl orange)	[69]
	Microparticles	In situ synthesis of TiO_2 NPs	Drug delivery systems	[70]
	Coating suspension	Dispersion of TiO_2 NPs into a CH solution with glycerol	Protective coatings for paper packaging	[71]
	Beads	Mixture of CH solution with TiO_2 NPs followed by coagulation	Photocatalysis (adsorptive abatement of aqueous arsenic)	[72]
	Membranes	Carboxymethyl CH modified TiO_2 NPs embedded into Pebax-1657 polymer matrix	CO_2 removal-air purification	[73]
	Films	Addition of colloidal TiO_2 to GO/CH nanometre film followed by freeze-drying	Packaging	[74]
	Films	Solvent casting of CH/TiO_2 NPs suspension	Packaging	[75]
	Coating	Dispersion of TiO_2 NPs in a CH solution followed by electrodeposition	Glucose biosensors	[76]
	Films	Dispersion of TiO_2 NPs and Ag NPs in a CH solution followed by casting	Antifouling materials	[77]
	NPs	Preparation of magnetic CH/EDTA NPs by W/O emulsion cross-linking followed by loading with TiO_2 NPs	Photocatalytic adsorbents	[78]

(continued)

Table 3.2 (continued)

Metal oxide	Type of material	Methodology	Application	References
	Nanofibrous scaffolds	Synthesis of GO/TiO$_2$/DOX composites, mixing with CH/PLA solution followed by electrospinning	Drug delivery systems	[79]
	Coating	In situ synthesis of TiO$_2$ NPs in a CH solution	Acetylcholinesterase biosensor for detection of pesticides	[80]
ZnO	Coatings	In situ synthesis of ZnO NPs	Cotton finishing	[81]
	Films	Mixing of CH/ZnO hybrid with a castor oil polymer solution followed by casting	Wound healing	[82]
	Beads	In situ synthesis of ZnO NPs	Photocatalysis decolorization	[83]
	Coating	Solvent casting of MWNT and ZnO NPs onto carbon electrodes followed by coating with CH	Sensors for determination of neurotransmitters	[84]
	Nanospheres	Nonsolvent-aided counter-ion complexation method	Cell imaging and cancer therapy	[85]
	NPs	Coating of ZnO NPs with carboxymethyl chitosan	Cancer therapy-delivery of curcumin	[86]
	Sponges	CH sponges impregnation with Ag/ZnO composite	Wound healing	[87]
	Film	Casting of a CH/CMC/ZnO NPs suspension	Packaging material (cheese)	[88]
	Film	Casting of a CMC/CH/ZnO NPs suspension	Packaging material (bread)	[89]
	Coating	In situ synthesis of ZnO meso crystals	Glucose biosensors	[90]
	Coating	Solvent casting	Anti-fouling coatings	[91]

(continued)

Table 3.2 (continued)

Metal oxide	Type of material	Methodology	Application	References
	Coating	Solvent casting of a CH/ZnO nanoneedles onto carbon electrodes	Sensor for determination of 4-nitrophenol	[92]
	Cross-linked hydrogels	Incorporation of ZnO NPs into a CH solution followed by cross-linking	Insulin drug delivery	[93]
	NPs	Ultrasonic sonication of CH NPs and ZnO NPs	Anti-fouling systems	[94]
SiO_2	NPs	Adsorption of insulin onto SiO_2 NPs followed by coating with CH	Drug carriers for insulin	[95]
	NPs	LbL assembly of CH and alginate onto SiO_2 NPs	Drug carriers for anticancer drugs	[96]
	Core-shell microspheres	Microfluidic approach	Catalysis	[97]
	NPs	Microemulsion method, in situ synthesis of SiO_2 NPs	Sensors	[98]
	NPs	Microemulsion method, in situ synthesis of SiO_2 NPs	Sensors	[99]
	Aerogels	In situ synthesis of SiO_2 NPs, followed by freeze drying	Scaffolds for tissue regeneration	[100]
	Nanogels	Self-assembly in a microemulsion system	Gene carrier material	[101]
	Particles	In situ Stober method	Scadium recovery	[102]
	NPs	Adsorption of CH-lactobioinic acid onto SiO_2 NPs	Drug carriers for anticancer drugs	[103]
	Aerogels	In situ synthesis of SiO_2 NPs, followed by freeze drying	Scaffolds for bone regeneration	[104]
	NPs	Adsorption of CH onto SiO_2 NPs	CO_2 adsorption	[105]

(continued)

Table 3.2 (continued)

Metal oxide	Type of material	Methodology	Application	References
	Films	In situ synthesis of SiO_2 followed by solvent casting	Food packaging	[106]
	Composites	Sol-gel method	Removal of dyes	[107]
	Xerogels	Sol-gel method followed by drying	Immobilization of proteins	[108]
	Particles	Covalent grafting of mesoporous SiO_2 with CH	Immobilization of proteins	[109]
Iron oxide	Films	Mixing of iron oxide NPs with a CH solution following by dip coating of FTO sheets	Apatsensor for determination of pesticides (malathion)	[110]
	NPs	Dispersion of Fe_3O_4 NPs in a CH solution	Glucose biosensor	[111]
	Microspheres	Mixing of Fe_3O_4 NPs, *Pseudomans* and CH solution followed by coagulation of the solution for the formation of the microspheres	Biocatalysts for biodiesel production	[112]
	NPs	Co-precipitation method (in situ synthesis of Fe_3O_4 NPs)	Catalysis, synthesis of spiroacridines	[112]
	NPs	Loading of chitosan NPs with iron oxide nanocubes	Cancer imaging	[113]
	NPs	Surface functionalization of dumdbell-like gol-iron oxide NPs with CH (water-in-oil emulsion)	Nanotheranostic systems	[114]
	Beads	In situ synthesis of iron oxide NPs	Sorbent for Cr(VI) removal	[115]
	Microspheres	Ion imprinting and inverse suspension cross-linking	Sorbent for selective removal of Cu(II)	[116]
	Microspheres	Mixture of CH solution and Fe_3O_4	Enrichment of intact proteins	[117]

(continued)

Table 3.2 (continued)

Metal oxide	Type of material	Methodology	Application	References
	NPs	Water-in-oil microemulsion methodology	Sorbents for mycotoxins (patulin)	[117]
Copper oxides CuO or Cu_2O	Microspheres	In situ synthesis of Cu_2O	Biosensor for Hg(II) detection	[118]
	Thin films	Mixture of CuO NPs with a CH solution followed by solvent casting	Photocatalysis	[119]
	Microparticles	In situ synthesis of Cu_2O	Photocatalysis	[120]
	Pellets	Mechanical mixing technique	Sensors for temperature and humidity	[121]
	Hydrogel	In situ synthesis of CuO NPs	Antibacterial hydrogel	[122]
	Films	Mixture of MMT/CuO with a CH solution followed by solvent casting	Antibacterial film	[123]
	Films	Mixture of CuO (or ZnO/CuO nanocomposites) NPs with a CH solution followed by solvent casting	Photocatalysis	[124]

Table 3.3 Examples of starch-based hybrids with metal oxides NPs, the preparation methodologies and potential applications

Metal oxide	Starch type	Methodology	Application	References
TiO_2	Soluble starch	In situ synthesis of TiO_2 nanostructures	Photocatalysis	[125]
	–	In situ synthesis	Photocatalysis	[126]
	Sodium carboxymethyl starch	Multi step procedure	Solar cells	[127]
	Starch aqueous solution	Post-modification of TiO_2 NPs with starch	Nanoadsorbents for metal ions absorption and preconcentration	[128]
ZnO	Rice starch	Solution casting	Shape memory packaging films	[129]
	Tapioca starch	Solution casting	Packaging films or coating	[130]
	Rice starch	Solution casting	Packaging films	[131]
SiO_2	Cassava starch	Melting mixing	–	[132]
	Hydroxypropyl starch	Solvent casting	Biodegradable food packaging films	[133]
	Modified starch and corn starch	Mixture of the different components	Binder for particle board panels	[134]
Iron oxide	–	Commercial starch coated magnetic iron oxide NPs were aminated and simultaneously PEGylated and heparinized	Magnetic imaging and tumour targeting	[135]
	–	Co-precipitation method	Sorbents for hevalent chromium removal	[136]
	–	Modification of hydrous ferric oxide NPs with starch via a water-based approach	Sorbents for arsenate removal	[137]
	–	Synthesis of Fe_3O_4 NPs, followed by adsorption of starch and grafting of poly(ethylene phthalate)	Drug delivery systems	[138]
	–	Co-precipitation method	Nanosystems for curcumin release	[139]

(continued)

Table 3.3 (continued)

Metal oxide	Starch type	Methodology	Application	References
	Commercial starch/polycaprolactone blend	Incorporation of commercial iron oxide NPs into a 3D structure of aligned starch/polycaprolactone fibres	Scaffolds for tendon regeneration	[140]
	Starch from rice	Alkaline oxidative hydrolysis followed by encapsulation with a starch-silica hybrid	Sorbents for pesticides removal	[141]
	Corn starch (27% amylose)	Adsorption of itaconate-functionalized starch onto the surface of iron oxide NPs followed by copolymerization with N-isopropylacrylamide	Theranostic applications	[142]
	–	In situ synthesis of iron oxide NPs	Nanosystems for curcumin release	[143]
	Short-chain amylose	Non-emulsion based approach, co-precipitation	Several biomedical applications	[144]
Cupper oxides CuO or Cu_2O	–	In situ synthesis of CuO NPs	Photocatalysis	[145]
	Potato starch	In situ synthesis of CuO NPs	Biomedical applications (anticancer activity)	[146]
	–	Blending of starch and CuO NPs with a polyol before the polyurethane synthesis	Antimicrobial foams for infections control	[147]

3.1 Cellulose/Metal Oxide NPs Hybrid Materials

Cellulose, including wood fibres, paper, cotton, regenerated cellulose and nanocellulose substrates, namely cellulose nanofibrils (CNFs), bacterial cellulose (BC),
cellulose nanocrystals (CNCs), and cellulose derivatives (mainly cellulose acetate
(CA) and carboxymethylcellulose (CMC)) have been extensively investigated in
combination with different metal oxides, originating a plethora of hybrid materials

in the form of films [13, 26, 27, 33, 37, 40, 65], membranes [12, 15–17, 21, 23, 25, 30, 31, 36, 43, 46, 48, 57, 63, 66, 67], aerogels [14, 24, 32, 47, 53, 148, 149], hydrogels [40, 62], nanoparticles [49–52, 54–56, 58, 59], and fibres (including cotton fabrics and paper sheets and filter paper) [11, 18, 22, 31, 38, 60], among others [19, 42, 45, 150] (Table 3.1).

For example, the combination of cellulose and TiO_2 NPs is one of the most studied strategies to fabricate bio-based materials for photocatalytic applications [11, 14–16, 18, 149]. As a representative study, Fujiwara et al. [18] recently reported the fabrication of a photocatalytic paper using TiO_2 NPs confined in hollow silica capsules. This strategy is particularly interesting because it allows the fabrication of photocatalytic active papers with photo-resistance against self-degradation of the cellulosic support. The obtained papers showed high photocatalytic activity (>35 $\mu mol\ h^{-1}$) in the photocatalytic decomposition of volatile organic compounds (VOCs), namely 2-propanol, being promising candidates materials for air cleaning.

Equally worth of reference is the work carried out by Li et al. [16] on the design of a laccase immobilized BC/TiO_2 functionalized membrane. Dye degradation experiments showed that under UV-irradiation the dye degradation efficiency increased and that these new hybrid BC membranes, with bio and photocatalytic functionalities, are valid for industrial degradation of textile dyes. In a distinct vein, nanostructured thin films containing TiO_2 and Au NPs supported in BC membranes were investigated as flexible photocatalytic devices to produce H_2 from an ethanol aqueous solution [15]. The nanostructured films were prepared by layer-by-layer assembly using polyelectrolytes and previously prepared TiO_2 and Au NPs. The best production of H_2, measured using gas chromatography, for the obtained films was of 0.70 mmol cm^{-2} when irradiated for 3 h.

Cellulose/TiO_2 hybrids have also been explored in the biomedical field [19–21], food packaging [13], as sensors [17] and nanosorbent systems [12], however in lower extent. As illustrative examples, the in vivo burn wound healing potential of BC/TiO_2 nanocomposites was investigated in burn wound model through wound area measurement, percent contraction and histopathology and the results pointed to a good healing pattern with 71% of wound contraction and formation of healthy granulation tissue [20]. In a different perspective, Pang et al. [17] reported an approach to prepare cellulose/TiO_2/polyaniline composite nanofibres for application as room temperature ammonia gas sensor. Briefly, electrospun cellulose acetate nanofibres were deacetylated to obtain regenerated cellulose nanofibres that were immersed into TiO_2 sol to produce cellulose/TiO_2 nanocomposites. Finally, in situ polymerization of aniline allowed to deposit polyaniline onto the surface of the nanocomposites. Evaluation of the gas sensing properties of the resulting hybrid nanofibres revealed that their response values and sensitivity (>5 for 250 ppm of ammonia) were much higher than those of cellulose/polyaniline counterparts (around 2).

In the latter years, a huge number of cellulose/ZnO hybrids have also been described in literature [151], with applications ranging from antibacterial materials for food packaging [22, 30], biomedical applications [23, 24, 31], water purification systems [24, 32], sensors [25, 26], photocatalysis [33, 35], UV-absorbing materials [152] and functional textiles [34], among others. Some other studies have their focus essentially on improving synthetic approaches to obtain this type of hybrids [153–156].

For instance, the self-assembly of hierarchically structured cellulose@ZnO hybrids in solid-liquid homogeneous phase was investigated as a strategy to produce antibacterial materials for food packaging [30]. The self-assembly mechanism was systematically studied, and it was found that the electrostatic attraction between cellulose and ZnO NPs was the driven force for the establishment of the first level structure. In a different study, Khalid et al. [31] fabricated antibacterial BC/ZnO hybrid materials for application as innovative dressing systems for burn wounds. The in vivo wound healing ability of the materials was investigated in burn BALB^c mice model and revealed a considerable activity (66%). Recently, Wang et al. [32] produced superfast adsorption-disinfection cryogels decorated with CNCs and ZnO nanorod clusters for water-purifying microdevices. These cryogels were obtained through a one-pot copolymerization of acrylamide and N',N''-methylenebis(acrylamide) monomer, 2-(dimethylamino)ethyl methacrylate monomer and flower like CNCs/ZnO nanohybrids and showed high mechanical strength, adsorption capacity of 30.8 g g^{-1}, superfast adsorption time (2.5 s), stable swelling/deswelling ability upon 10 cycles and dual temperature/pH responsiveness. Equally interesting, Mun et al. [26] reported a flexible and disposable cellulose/ZnO hybrid film, for conductometric glucose biosensing. The hybrid film was produced by blending ZnO NPs with a cellulose solution in lithium chloride/N,N-dimethylacetamide, followed by curing in an isopropyl alcohol/water mixture. The biosensor was finally obtained by physical adsorption of glucose oxidase into the film. It was observed that the enzyme activity increased with the increase of the ZnO weight ratio in the film and that the film can detect glucose in the range 1–12 mM. Another representative study investigated the degradation of organic contaminants, (Congo red and Reactive yellow-105) using a cellulose acetate-polystyrene membrane impregnated with ZnO NPs under sunlight irradiation [35]. It was concluded that the addition of ZnO NPs to the membrane decrease its fouling and improve the permeation quality with above 90% of photocatalytic degradation efficiency for the studied dyes. Moreover, the reusability of the hybrid membranes was studied, and no significant changes were perceived until four cycles.

Cellulose/SiO$_2$ hybrids have similarly been widely explored in the latter years, with applications in vast domains, including insulation [41, 44, 47] and flame-retardant [45, 148] materials, sensors [36, 39, 40], water purifications systems [48], separation [38, 46] and packaging [37] materials, functional textiles [150] (as recently reviewed) drug delivery [42], among many others [43]. As an illustrative example, transparent cellulose-silica aerogels with excellent flame retardancy were prepared by in situ formation of SiO$_2$ NPs via a two-step sol-gel process in a cellulose gel [148]. The results showed that the increasing of the silica content increases the transparency, compressive, thermal and thermo-oxidative properties of the obtained aerogels. For an aerogel with 33.6% of silica, the transmittance of the composite at 800 nm was of 78.4%. In a different study, Kim et al. [38] developed porous structure-tuned cellulose nanofibre paper separators, following an architectural methodology based on SiO$_2$ NPs, for lithium ion batteries. The porous structure of these cellulose separators can be tuned by varying the amount of SiO$_2$ NPs in the cellulose nanofibres suspension, with the one with 5% showing the highest ionic conductivity. Equally interesting, Evans et al. [39] produced a silica NPs-modified microfluidic analytical device for enzymatic reactions with clinical significance. The devices were fabricated in filter paper and using a CO_2 laser engraver. The addition of silica NPs promoted an increment of the colour intensity and uniformity. The obtained cellulose devices allowed the detection of three analytes (glucose, glutamate and lactate) in artificial urine samples with detection limits of 0.50, 0.25 and 0.63 mM, respectively. Recently, Albertini et al. [48] described novel boron-chelating membranes based in hybrid mesoporous silica NPs immobilized in a cellulose acetate membrane for water purification. These membranes showed boron removal efficiencies of up to 93% and can be used in multistage filtering systems with continuous operation. In a completely different vein, Hakeem et al. [42] designed innovative cellulose conjugated silica NPs based stimuli-responsive nanosystems for cancer treatment. Esterification between cellulose and mesoporous silica NPs was carried out to conjugate cellulose on the surface of the NPs aiming to control the release (avoid premature release) of doxorubicin (DOX) under physiological conditions. DOX release from the cellulose conjugate silica NPs was only 10.9% at pH 7.4 in comparison with 75.4% from pure silica NPs.

Because of the well-recognized magnetic properties of some iron oxides, in the latter years, the combination of cellulose and iron oxide NPs have been essentially investigated to produce magnetic materials for application in water purification [50, 53, 56], drug delivery [51, 54, 58], magnetic hyperthermia treatments [54, 55], imaging [52], sensing [57, 60] and proteins separation [49, 59]. Other less reported applications of cellulose/iron oxides include magneto-optical components [157], magneto-responsive nanofillers [158], shielding materials [159], wound healing membranes [160], multifunctional pigments [161] and catalysts [162, 163], among others. It is important to emphasize that most of above mentioned studies are related with the use of cellulose, and cellulose derivatives, as capping agents to produce more stable composite iron oxide NPs rather than to take advantage of the supramolecular structure of cellulose substrates.

For example, Luo et al. [56] fabricated magnetic cellulose beads with micro and nanoporous structures following an extrusion dropping technology from NaOH/urea aqueous solution (Fig. 3.1). The hybrid beads incorporated with carboxyl modified magnetite NPs and nitric acid modified carbon nanotubes showed sensitive magnetic response and efficient removal performance for several heavy metal ions from water (maximum adsorption amounts calculated of 47.64, 37.99 and 22.30 mg g^{-1} for Cu^{2+}, Pb^{2+} and Zn^{2+}, respectively).

In a distinct study, Chen et al. [50] reported the design of β-cyclodextrin-modified cellulose nanocrystals superparamagnetic nanorods for removal of pharmaceutical residues. These functional cellulose nanorods were obtained by grafting of β-cyclodextrin onto the surface of $Fe_3O_4@SiO_2$ hybrids and showed good adsorption of two model compounds, namely procaine hydrochloride (13.0 mg g^{-1}) and imipramine hydrochloride (14.8 mg g^{-1}). Another interesting study focused on the development of superparamagnetic iron oxide NPs for targeted delivery of curcumin and hyperthermia treatment [54]. $MnFe_2O_3$ NPs (15–20 nm) were used as core

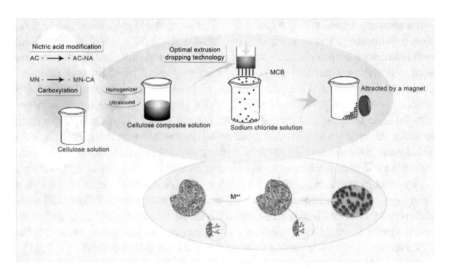

Fig. 3.1 Preparation of magnetic cellulose-beads and the absorption mechanisms of heavy metal by the beads. Reprinted with permission from [56]. Copyright 2016 American Chemical Society

materials and modified with carboxymethyl cellulose, folic acid and a dual responsive polymer by microwave irradiation. The obtained nanosystems are not harmful for normal cells and cancer cells but show high drug loading (89%) and fast drug delivery at pH 5.5. Equally stimulating, Darwish et al. [55] developed a new approach to produce magnetic NPs with shells of oleic acid, polyethyleneimine, and polyethyleneimine-methyl cellulose. The obtained NPs display antibacterial activity against both *Staphylococcus aureus* and *Escherichia coli*, specifically 10% growth inhibition (EC10) for concentrations <150 mg L^{-1}, and high specific adsorption rate being effective for high heat-generation in alternating magnetic fields and therefore for magnetic hyperthermia cancer treatment and simultaneous decrease of bacterial infectious co-localized with cancers.

Another interesting study involved the development of magnetic cellulose nanocrystals/Fe$_3$O$_4$ NPs hybrids, complexed with Cu (II) ions, to be used as regenerable, efficient and selective platforms for protein removal [59]. The performance of these hybrids was assessed by selectively removing lysozyme from aqueous media (binding capacity close to 860 mg g^{-1}) and nearly full protein recovery (~98%), and then confirmed in the separation of that protein from egg white.

In a different vein, an inexpensive methodology to deposit magnetite NPs on cellulose filter paper sheets and to conjugate these NPs with anti-human-IgM antibodies using polydopamine as linker was developed [60]. These novel magnetic papers were tested as solid immunoassay platforms for IgM-dengue antibodies recognition and showed improved analytical response, precisely two orders more sensitive than the traditional ELISA systems.

Compared to other metal oxide NPs, the literature available on cellulose/copper oxide hybrid materials is scarcer and essentially centred on the production of antibacterial materials for different applications including, food packaging [65, 67], wound dressing [62], water purification [63, 67] and functional textiles [64]. Cellulose/CuO hybrids for catalysis [61] and supercapacitors have also been reported in the latter years.

In a representative example, the in situ formation of CuO NPs inside a carboxymethylcellulose-based hydrogel matrix attached to cotton fabrics was investigated [62]. The performance of the CuO hydrogels as potential wound dressing was tested for antibacterial activity against *E. coli, Pseudomonas aeruginosa, S. aureus* and *Bacillus subtilis*, and showed high antibacterial activity (19, 19, 20 and 10 mm cm^{-1} of inhibition zone diameter). In a different study, Booshehri et al. [63] reported also a simple in situ method for the deposition of CuO NPs onto cellulose paper and their superior antibacterial activity against *E. coli* and *E. aureus* (100% CFU reduction after 150 and 180 min, respectively) was attributed to the release of copper ions from CuO NPs. Based on these results the authors proposed their use as cheap systems for drinking water purification. In a completely different way, Peng et al. [66] prepared bacterial cellulose membranes coated with polypyrrole and CuO NPs for application as flexible electrodes in supercapacitors. Electrochemical measurements showed that the supercapacitors employing the polypyrrole/CuO/BC electrodes had a specific capacitance of 601 F g^{-1} with an energy density of 48.2 W kg^{-1} and a power density of 85.8 W kg^{-1} at a current density of 0.8 mA cm^{-2}.

3.2 Chitin/Metal Oxide NPs Hybrid Materials

Because of the low solubility of chitin in common solvents, as well as due to the interesting properties of its main derivative chitosan, only a few examples of chitin/metal oxide hybrid materials were reported in literature in the last 5 years. For example, Wysokowski et al. [164] reported a hydrothermal synthesis of β-chitin/ZnO nanostructured hybrids with antibacterial properties against Gram positive bacteria that can be employed for the preparation of wound dressing materials. In a similar vein, β-chitin hydrogel/ZnO bandages were fabricated and tested as an alternative to existing wound healing bandages [165]. ZnO NPs were added to β-chitin hydrogels and the porous bandages were finally obtained by freeze-drying. The obtained hybrids showed controlled swelling and degradation as well as higher blood clotting and platelet activation, when compared with the control, resulting in a faster healing and collagen deposition ability. Recently, Sahraee et al. [166] fabricated gelatin films containing ZnO NPs and chitin nanofibres by solvent casting of aqueous suspensions containing different amounts of ZnO (0, 1, 3 and 5%) and chitin (0, 1, 3, 5 and 10% relative to gelatin). The addition of chitin and ZnO NPs increased the thermal stability and antifungal activity of the films, and the incorporation of ZnO improved also their water vapour permeability (WVP) and mechanical properties. Therefore, these hybrid films can be used as active food packaging systems. In a similar way, chitin nanowhiskers based hybrids with ZnO and Ag NPs were used for the preparation of multifunctional carboxymethyl cellulose films with improved mechanical properties [167]. The CMC film with 5 wt% of this hybrid was homogeneous and showed the highest UV-barrier properties and strong antibacterial activity against *E. coli* and *Listeria monocytogenes*.

Chitin/silica hybrids have also been investigated, however in a much more fundamental manner. Bashenov et al. [168] prepared monolithic silica-chitin materials under extreme biomimetic conditions (80 °C and pH 1.5) using 3D chitinous matrices and studied them by different techniques to establish a mechanism for the chitin-silica interactions. In another study, chitin/silica hybrid thin films were prepared by colloidal self-assembly of chitin nanorods and siloxane oligomers and studied for the first time by PeakForce QNM AFM mode to explore their structure and mechanical properties [169].

In a distinct vein, Ramos et al. [170] investigated novel chitin hydrogels reinforced with TiO$_2$ NPs for removal of arsenic from natural waters. The equilibrium assay showed a maximum adsorption capacity of 3.1 mg g^{-1} and the reutilization of the hybrid gel sorbent was assessed up to five cycles.

In another representative work, Wang et al. [171] fabricated a sunlight photocatalyst by in situ synthesis of Cu$_2$O in regenerated chitin/graphene oxide films, where the porous chitin film was used as the micro-reactor for the formation of the Cu$_2$O NPs. It was also found that the chitin film could load both Cu$_2$O and graphene, controlling the size of the copper oxide NPs and leading to easy recycle and reuse of the photocatalyst.

3.3 Chitosan/Metal Oxide NPs Hybrid Materials

Chitosan (CH), the deacetylated derivative of chitin, because of its recognised antimicrobial properties, solubility in acidic aqueous solutions and film-forming ability has been extensively investigated in combination with different metal oxides to produce panoply of hybrids, in the form of films [77, 82, 88, 89, 106, 110, 124], (nano)particles [74, 85, 86, 94–96, 98, 99, 102, 103, 109, 111–113, 172, 173], aerogels [69, 87, 100, 104], microspheres [97, 112, 116–118], nanogels [101, 122], beads [72, 83, 93, 115], nanocomposites [70, 107, 174], xerogels [108], coatings [68, 71, 76, 81, 91, 92], among others [121], as summarized in Table 3.2.

CH/TiO$_2$ hybrids are one of the most investigated CH based hybrid classes, with a huge number of studies focusing on the exploitation of their adsorbing and photocatalytic properties to obtain materials for environmental remediation [68, 69, 72, 78], packaging [71, 74] and antifouling materials [77]. For example, novel core-shell surface imprinted chitosan-TiO$_2$ hybrids for pollutants removal were developed by combination of molecular imprinting, chitosan sorption capacity and nano TiO$_2$ photocatalytic technology, using methyl orange as the template [69]. As expected, the composites showed improved photocatalytic selectivity for methyl orange, when compared with the non-imprinted counterpart, because of the existing of proper locations generated by this technique. It was also found that the removal of methyl orange was essentially due to photodegradation rather than adsorption and that the nanosystems could be reused for 10 cycles without the need for desorption and regeneration.

In a different study, a bio-photocatalyst for arsenic diminution in aqueous media was fabricated by immobilization of TiO_2 NPs and feldspar particles in a chitosan matrix [72]. Contrarily to other TiO_2 based remediation strategies, these materials effectively removed arsenic in a wide range of pH values. UV irradiation improved removal efficiencies from 33 to 73% and from 23 to 84% for arsenate and arsenite, respectively, for an initial concentration of 4800 $\mu g \, L^{-1}$. The enhanced removal of As under UV-irradiation was associated by the UV cleavage of CH polymeric chain releasing new carboxyl groups.

Recently, Alizadeh et al. [78] reported novel cross-linked magnetic EDTA/TiO_2/CH nanosystems for removal of Cd (II) and phenol from aqueous solutions. Results revealed that the hybrids with an average diameter of 40 nm had the best performance in the adsorption of Cd (II), and on the photodegradation of phenol at a pH of 5–6. In this case, the removal of the contaminants takes advantage of both the adsorption and photocatalytic activity of the hybrid. The maximum adsorption capacity of Cd (II) predicted from Langmuir model was *ca.* 209 mg g^{-1} and phenol degradation efficiency was up to 90%. Regarding packaging applications of CH/TiO_2 hybrids, for instance, Tang et al. [71] reported a CH/TiO_2 coating for packaging materials. The obtained results showed that the increasing of the TiO_2 NPs loading (from 0 to 20%) decrease the viscosity and dynamic viscoelasticity of the coating suspensions, that facilitates the production of high solid content coatings, and enhanced the antibacterial activity (against *E. coli*) and mechanical performance of surface coated cellulosic materials, with an optimal TiO_2 loading of 10%. In a different study chitosan/TiO_2 NPs films with efficient antibacterial activity (100% sterilization after 12 h) against food-borne pathogens (*E. coli, S. aureus, Candida albicans* and *Aspergillus niger*) were obtained by solvent casting [75]. TiO_2 addition led to higher hydrophilicity, improved mechanical properties, and decrease light transmittance of the films in the visible region. The films were further tested for packaging of red grapes showing positive results (preservation of grapes lasted for 15 days with pure chitosan film and 22 days with the CH/TiO_2 film). Natarajan et al. [77] investigated the antifouling activity of CH/TiO_2/Ag NPs films against fresh water algae (*Scenedesmus* sp. and *Chlorella* sp.) under different photo conditions. The results showed that *Scenedesmus* sp. is more sensitive to all films investigated than *Chlorella* sp. under both UV-C exposure and dark conditions. The slime formation, biomass, lipid peroxidation and uptake of NPs correlated well with toxicity of the algae, indicating that the species variations considerably influences the antifouling action of all tested films, including those with TiO_2 NPs.

Other applications of CH/TiO_2 NPs hybrid materials include CO_2 removal [73], drug delivery [70] and biosensors [76, 80], among others [175–177]. In an illustrative example, efficient CO_2 removal using mixed matrix membranes with carboxymethyl chitosan (or 3-aminopropyl-diethoxymethylsilane) modified TiO_2 NPs was described [73]. TiO_2 NPs modification with carboxymethyl chitosan and surface grafting with 3-aminopropyl-diethoxymethylsilane led to an increase of 50–66% in the CO_2 permeability and of 30–33% in CO_2/N_2 selectivity when compared with the pristine polymeric matrix. The performance of these membranes is satisfactorily above the 2008 Robeson upper bond and showed acceptable durability during the 36 h experiments. In a distinct mood, Kamari and Ghiaci [70] produced ibuprofen/modified-CH/TiO_2 NPs hybrid drug delivery systems. The coating of the chitosan-ibuprofen composite with TiO_2 enhanced the drug entrapment and reduced the drug release from 24 to 54 h. Moreover, these hybrids showed a pH sensitive drug release profile, with slower release in the simulated gastric fluid (pH 1.2). These results demonstrate that these systems could decrease the side effects of ibuprofen by minimizing its release in the stomach.

Regarding the design of CH/TiO_2 biosensors, as an expressive example, Cui et al. [80] developed a highly stable acetylcholinesterase biosensor based on CH/TiO_2/graphene hybrids for detection of organophosphate pesticides. The fabrication process was based on the adsorption of acetylcholinesterase on a CH/TiO_2 sol-gel and reduced graphene oxide multi-layered mesoporous matrix. The catalytic activity of glassy carbon electrodes coated with this hybrid to acetylcholine was significantly higher than those of the individual components in the matrix. The linear detection of this biosensor to a model organophosphate compound ranged from 0.036 to 22.6 µM with a detection limit of 29 nM and detection time of 25 min. Moreover, it was found that the biosensor was stable under detection and storage (30 days).

Similarly to other polysaccharides, CH based ZnO NPs hybrids have also been extensively investigated in the last 5 years. Most of the applications of CH/ZnO NPs based materials take advantage of the well-known antimicrobial activity of both ZnO NPs and CH, as for example, wound healing materials [82, 87, 178], anti-biofouling systems [91, 94], packaging materials [88, 89] and antibacterial cotton finishing [81]. Other applications of CH/ZnO hybrids include photocatalysis [83], sensors [84, 90, 92], and nanosytems for cancer therapy [85, 86], imaging [86] and drug delivery [93].

For instance, Díez-Pascual and Díez-Vicente [82] prepared biodegradable composites based on castor oil polymeric films reinforced with CH/ZnO NPs via solution mixing and solvent casting for application as wound healing materials. Several properties of the films, namely thermal stability, mechanical performance, degree of porosity, water absorption, water vapor transmission, oxygen permeability and biodegradability increased with the chitosan/ZnO hybrid loading. The antibacterial activity of the films against *E. coli*, *S. aureus* and *Micrococcus luteus* was investigated in the absence and presence of light showing an increase biocide effect with the content of the hybrid and stronger against Gram-positive bacterial strains. Moreover, hybrids with less than 5.0% of ZnO NPs showed very good cytocompatibility and promoted faster wound healing than neat castor oil films and control gauze. In another study [87], sponge-like Ag/ZnO nanocomposite loaded chitosan dressings were fabricated by lyophilisation process followed by incorporation of the Ag/ZnO nanocomposite into the chitosan porous material. Results showed that the prepared dressings have high porosity (81–88%) and swelling (21–24 swelling ratio), enhanced antibacterial activity (against drug-resistant pathogenic bacteria) and blood coagulation and low in vitro cytotoxicity. In vivo studies in mice reveal that these porous dressing augment the wound healing and endorse re-epithelialization and collagen deposition.

In a totally different vein, composite packaging materials prepared using CH, CMC and ZnO NPs were prepared by a casting method and characterized by different techniques to evaluate their potential to enhance the shelf life of Egyptian soft white cheese [88]. The films demonstrated good antibacterial activity against Gram positive (*S. aureus*) and Gram negative (*P. aeruginosa* and *E. coli*) bacteria and fungi (*C. albicans*), and consequently increased the cheese shell life (stored at 7 °C). In a comparable fashion, Noshirvani et al. [89] also used CMC/CH/ZnO hybrid films to increase the self-life of bread from 3 to 35 days.

In another recent contribution, Rasool et al. [94] described highly stable green biocide formulations, composed of ZnO-interlinked CH NPs, for inhibition of film formation caused by sulphate reducing bacteria in inject seawater. The inhibition of these bacteria was found to be concentration dependent with more than 73% inhibition at 250 mg mL^{-1} of chitosan/ZnO NPs at an initial ZnO loading of 10%. Moreover, the cytotoxicity and environmental impact of these nanosystems was evaluated by using zebrafish embryos and neither mortality nor teratogenic effects were observed using the acute toxicity assay.

CH/ZnO NPs based materials have also been considerably explored for the development of sensors for different target molecules because of their intrinsic properties. For example, Wang et al. [84] reported a disposable electrochemical sensor for simultaneous determination of norepinephrine and serotine in rat cerebrospinal fluid composed of multi-walled carbon nanotubes (MWNTs)-ZnO/CH hybrids modified screen-printed electrodes. The obtained results showed that the responses of both neurotransmitters significantly improved due to the high catalytic activity of the hybrids. Moreover, the peak currents of norepinephrine and serotine were linearly dependent on their concentrations in the ranges 0.5–30 and 0.05–1 μM, with detection limits of 0.2 and 0.01 μM, respectively, and the electrodes can be stored at 4 °C for at least 3 months. In a similar vein, a highly sensitive electrochemical method for the determination of 4-nitrophenol in water based on a chitosan crafted ZnO needles modified screen printed carbon electrode was developed [92]. Under optimum conditions, the modified electrode showed a wide linear response, from 0.5 to 400.6 μM towards the detection of the target molecule, with a detection limit of 0.23 μM. these electrodes showed also a long-term stability.

Finally, CH/ZnO NPs hybrids also showed enormous potential as cancer imaging and therapy platforms and drug delivery systems. For example, Zhao et al. [85], reported novel CH/ZnO quantum dots-based NPs (around 100 nm) with enhanced photoluminescence stability and longer retention time in cells, compared with pure ZnO quantum dots. Moreover, these NPs can also be loaded with drugs, as DOX, without affecting the fluorescent properties of the ZnO quantum dots allowing the simultaneous cell imaging and cancer therapy. In another study [93], smart and flexible CH/ZnO hydrogels were prepared and investigated as nanosystems for insulin nasal cavity drug delivery using in vivo rat models. The formulation of the hydrogels decreased the blood glucose concentration (50–65% of the initial glucose concentration) for at least 4–5.5 h after the administration and have not shown cytotoxicity.

CH/SiO$_2$ NPs hybrids have also been extensively investigated in different applications, including packaging films [106], drug delivery systems [95, 96, 103], gene carriers [101], hybrid scaffolds [100, 104], catalysis [97], adsorbers [102, 105, 107], protein immobilization [108, 109] and sensors [98, 99].

As an illustrative example, Yu et al. [106] developed biodegradable PVA/CH films with improved mechanical properties by in situ incorporation of SiO$_2$ NPs. The tensile strength of the films improved 45% when only 0.6% of SiO$_2$ was incorporated. Moreover, the oxygen and moisture permeability of the hybrid films decreased 25.6 and 10.2%, respectively, making them interesting candidates for food packages with extended preservation time.

In a totally different vein, mesoporous silica NPs coated with low molecular weight CH were investigated as injectable controlled release carriers for insulin [95]. Results of surface tension measurements showed that insulin was absorbed into the silica NPs and interacted with CH. Moreover, their stability was confirmed by different analytical techniques and in vivo results pointed for an insulin prolonged released after loading in the silica NPs. In another interesting example, Feng et al. [96] proposed a layer-by-layer (LbL) approach to prepare pH-responsive alginate/CH multi-layered coated silica NPs drug delivery systems with improved efficacy and blood compatibility. As a proof-of-concept, the nanocarriers were loaded with DOX and tested both in vitro and in vivo as pH responsive systems showing a sustained intracellular DOX release, a prolonged accumulation in the nucleus and a longer systemic circulation time and slower plasma elimination rate when compared with free DOX. In another contribution, a pH sensitive system based on lactobionic acid modified CH-conjugated mesoporous silica NPs for co-delivery of sorafenib, a multi-tyrosine inhibitor, and ursolic acid, a sensitive agent for sorafenib, was developed (Fig. 3.2) [103]. These nanosystems showed synergetic cytotoxicity and attenuated the adhesion and migration of asialoglycoprotein receptor over-expressing liver cancer SMMC-7721 cells at non-cytotoxic concentrations. In addition, this nanocomplex considerably enlarged the cellular apoptosis and down-regulated the proteins related with cell proliferation and tumour angiogenesis. Overall, the co-delivery of ursolic and sorafenib by pH-sensitive chitosan-conjugated mesoporous silica NPs represents a promising strategy for hepatocellular carcinoma combinational treatment, in particularly for metastasis chemoprevention.

In the latter years, CH/SiO$_2$ hybrid materials have also been explored as scaffolds for tissue regeneration. For example, Wang et al. [100] reported highly flexible CH/silica hybrid scaffolds with oriented pore structures produced by a sol-gel method with a unidirectional freeze casting method. Structural characterization of the hybrids and dissolution assays indicated the covalent cross-linking of CH and the silica network. The scaffolds showed directional lamellar and cellular morphologies along and perpendicular to the freezing directions, respectively. Compression tests results showed that the hybrids with 60% CH were flexible and elastomeric perpendicularly to the freeze direction whilst being elastic-brittle parallel to the freeze direction (with a compressive strength of about 160 kPa) showing enormous potential for surgical implants by simply squeezing and pressing the scaffold from the perpendicular direction into the tissue imperfection without affecting the mechanical properties of

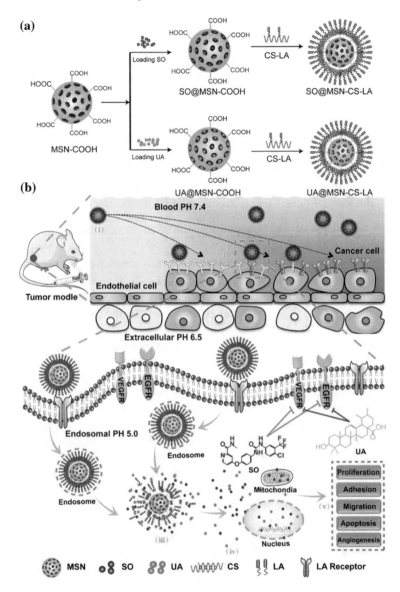

Fig. 3.2 Formation of lactobionic acid modified pH-sensitive chitosan-conjugated mesoporous silica NPs loaded with ursolic acid and sorafenib and the co-delivery of the bioactive molecules. Reprinted with permission from [103]. Copyright 2017 Elsevier

the stronger side. In a more recent study, similar CH/silica scaffolds were prepared and investigated on guided bone regeneration [104]. In vitro studies showed that the scaffolds promoted a fast formation of calcium phosphate mineral without cytotoxic effects, against murine macrophage and endothelial cells, showing a bone bioactivity superior to the one of pure CH scaffolds.

In a totally different vein, Zhao et al. [97] described a facile microfluidic approach to fabricate CH/silica core-shell hybrids microspheres with effective protection of $-NH_2$ groups and improved mechanical performance (2 times) when compared with CH microspheres. When loaded with Cu (I) salts, these chitosan/silica systems are outstanding reusable catalysts for azide-alkyne cycloaddition reaction at room temperature. In this study, silica was not obtained in the form of NPs but as a shell coating.

CH/silica NPs hybrids show an enormous potentiality as absorbers. In an illustrative example, mesoporous CH/SiO_2 particles (211 nm) were synthesized and evaluated for the absorption of CO_2 [105]. The BET (Brunauer–Emmett–Teller) surface area and the porous volume of these particles were found to be of 621 and 0.71 m^3 g^{-1}, respectively. The CO_2 adsorption was investigated by a volumetric methodology and the CH/SiO_2 hybrid particles showed a maximum CO_2 adsorption of 4.39 mmol g^{-1} at 25 °C and a high separation selectivity for CO_2-over-N_2 ($S_{CO_2/N_2} = 15.46$). In a very recent study, CH/silica hybrids, obtained by different techniques (physical adsorption and sol-gel method), were investigated for the removal of azo dyes from aqueous solutions [107]. For example, it was found that the sol-gel method allows to obtain hybrids with high CH content (16.7%) and specific area of 600 m^2 g^{-1}. Adsorption studies revealed the ability of all these materials to adsorb dyes, however, their adsorption capacities were differentiated, depending on the methodology, because of their distinct physicochemical, surface and structure properties.

In another interesting approach, a nanosensor based on CH/silica NPs doped with Ru(2,2'-bipyridyl)$_3^{2+}$ and fluorescein isothiocyanate, was used as surface paper pH indicator [99]. The nanosensor showed a linear dynamic range from 5.5 to 8.0. It was also verified its suitability for in situ surface pH sensor of paper fibre because it could infiltrate into the surface of the paper, as demonstrated with paper art samples.

In the latter years, the combination of CH and iron oxide NPs has also been considerable exploited, with focus on the design of magnetic sorbents for removal of metal ions from water [115, 116, 174] as well as proteins or other biomolecules from distinct aqueous systems [117]. For instance, Kong et al. [116] prepared magnetic copper imprinted CH/graphene oxide (GO)/iron oxide NPs by combination of ion imprinting and inverse suspension cross-linking. High adsorption capacity for copper was observe, particularly for the systems with graphene oxide. The maximum adsorption capacity was 132 mg g^{-1} at pH 6. The adsorption process follows the Freundlich isotherm equation and a pseudo-second order kinetic model. These hybrid beads showed a good stability and regeneration ability (up to five cycles adsorption-desorption). In another contribution, CH/iron oxide hybrids were obtained by one-pot green route and tested for sequestration of Pb (II) and Cd (II) from synthetic and industrial wastewaters [174]. In this case, the experimental results fitted better with the Langmuir isotherm and a pseudo-second order kinetic model. The maximum monolayer sorption capacities of Pb (II) and Cd (II) were found to be around 215 and 204 mg g^{-1}, respectively. These hybrid materials showed also excellent recyclability up to 5 cycles.

With a different purpose, CH coated Fe_3O_4 NPs were prepared by reverse oil-in-water microemulsion method and used as magnetic adsorbents for the adsorption of the mycotoxin patulin in juice-pH simulation systems [172]. The maximum adsorption capacity was of 6.67 mg g^{-1} within 5 days by adding 300 μg of the NPs into 10 mL of aqueous patulin 200 μg L^{-1} demonstrating their potential for the removal of this mycotoxin from industrial fruit juices.

Other interesting applications of CH/iron oxide-based NPs materials include sensors [110, 111], catalysis [112], cancer imaging [113] and nanotheranostic systems [114], among others [179, 180]. In a more fundamental mood, the thermal and magnetic properties of chitosan/iron oxide NPs [181] and the in vitro toxicity assessment of chitosan oligosaccharide coated iron oxide NPs [182] were investigated.

In an illustrative example, an electrochemical aptasensor based on CH/iron oxide film deposited on fluorine tin oxide (FTO) was fabricated for the detection of the pesticide malathion [110]. The biotinylated DNA aptamer sequence specific to this pesticide was immobilized onto the iron oxide/CH/FTO electrode by linking using streptavidin. Compared to enzyme-based sensors, the developed aptalectrode exploit the remarkable specific recognition elements for the detection of malathion. The obtained results showed that the electrodes exhibited a limit of detection of ~0.001 ng mL^{-1} within 15 min. Moreover, spike-in studies revealed a recovery of malathion in the range of 80–92% from lettuce leaves and soil samples demonstrating its potentialities. In a distinct vein, Chen et al. [112] developed magnetic whole-cell biocatalysts for enzymatic production of biodiesel from soybean oil, by immobilization of *Pseudomonas mendocina* cells into Fe_3O_4/CH microsheperes. A biodiesel production yield of 87.32% was achieved under the optimum experimental conditions (biocatalytic concentration of 10 wt%, water content of 10 wt%, 35 °C, methanol to water ratio of 4:1, and a four step addition of methanol). The biocatalyst has a very good reusability with a biodiesel production of *ca.* 84% after 10 cycles and could be recycled easily because of its superparamagnetic properties, demonstrating its potential for large-scale biodiesel industry. As an example of the exploitation of CH/iron oxide hybrids in cancer treatment and diagnosis, Kostevsek et al. [114] described the synthesis of CH based nanotheranostic systems following one-step synthesis and functionalization of dumbbell-like gold-iron oxide NPs. The results demonstrated the potential of these NPs as a photoacoustic imaging contrast agent in the phantom and as nano-heater for photothermal therapy.

CH/copper oxide hybrids were fairly investigated in the later years, particularly in combination with other metal oxide NPs, for application essentially in catalysis [119, 120], sensors [118, 121] and antibacterial materials [122, 123]. For example, thin films of CH and ZnO/CuO nanocomposites were prepared by solvent casting methodology and tested for the photodegradation of Fast Green Dye under artificial and solar irradiation [124]. The obtained results indicated that the best catalytic results were obtained with the nanocomposite in combination with UV light irradiation (with more than 90% of photodegradation). The study also concluded that the film was easy to separate without the need of a centrifuge or magnet as in several other studies. Chani [121] reported the synthesis of a hybrid composed of CH and a CuO-Fe_2O_4 nanopowder with a particle size of about 30 nm that were used to fabricate sensor pellets (10 mm i.d., thickness of 1 mm) for impedimetric sensing of temperature and humidity. The sensors perform best in the 20–80 °C temperature range but cover the entire room humidity range. In a distinct vein, a series of antibacterial carboxymethyl CH/CuO hybrid hydrogels were obtained by treating carboxymethyl chitosan with a Cu (II) chloride solution followed by oxidation [122]. SEM micrographs of the hydrogel revealed a uniform distribution of the CuO NPs (20–50 nm) within the CH hydrogel matrix. The obtained hydrogels showed rather higher swelling ability compared with the pure carboxymethyl chitosan hydrogels and excellent antibacterial activity against *S. aureus* and *E. coli*. Therefore, these hydrogels can find application in numerous biomedical fields where the control of bacterial infections is an important issue.

3.4 Starch/Metal Oxide NPs Hybrid Materials

In the latter years, the partnership between starch and metal oxide NPs (Table 3.3) has been essentially centred on the use of starch as reducing and stabilizing agent and/or capping biopolymeric layer on the synthesis of stable nanocomposite metal oxide NPs for distinct applications [125, 126, 128, 135–137, 139, 141, 142, 145, 146]. The studies dealing with the development of materials are almost limited to the preparation of nanocomposites (essentially films) [130–133, 138, 140, 144, 183], foams [134, 147] and gels [127].

The combination of starch with TiO$_2$ NPs have been used to prepare hybrids for application in packaging [183], photocatalysis [125, 126], solar cells [127] and sorbents for water purification [128]. Bao et al. [125] investigated the synthesis of TiO$_2$ nanostructures using other saccharides (β-cyclodextrin, chitosan and starch) as templates to tailor their morphology, crystal phase and photocatalytic activity. Remarkably, it was observed that saccharides not only direct the formation of the desired architectural crystals but also considerable influence the crystal phase. Rutile TiO$_2$ is obtained when using CH and β-cyclodextrin, while anatase TiO$_2$ is formed when using soluble starch. All hybrid nanomaterials show higher photocatalytic activity (when tested on a model dye) than those obtained without the templates. Another fascinating example involves a novel gel electrolyte produced by solidifying conventional polysulfide aqueous solution with sodium carboxymethyl starch as gelator [127]. The obtained gels exhibit high conductivity and beneficial ion transport due to theirs high water absorbing and holding capacity as well as porous network. These gel electrolytes could form a passivation layer coated on the surface of quantum dots/TiO$_2$ yielding quasi-solid-sate quantum dot-sensitized solar cells with a photoelectric conversion efficiency of 6.32%. In a different vein, Baysal et al. [128] reported starch coated TiO$_2$ NPs as innovative nanosorbents as a fast easy handled and environmentally friendly approach to separate and preconcentrate several heavy metals and for their analysis using a graphite furnace atomic absorption spectrometry. The detection limits achieved were 0.05, 0.28, 1.90 and 0.11 mg L^{-1} for cadmium, cobalt, copper, nickel and lead.

Starch/ZnO hybrids have also been considerably investigated in the latter years with emphasis on the development of nanomaterials for application in active packaging [129–131]. In an illustrative example, biodegradable films of starch/lysine and ZnO NPs with shape memory were produced by solvent casting methodology. The amounts of ZnO NPs was varied from 0 to 5% and it was observed that the increment of the ZnO content improved the thermal stability, decrease the solubility of the film in water, and increased the mechanical properties. In addition, the films showed shape memory properties when placed at room temperature and then at 55 °C. More recently, Marvizadeh et al. [130] developed a nano-packaging material for using as coating or edible film composed of tapioca starch, bovine gelatin and nanorod zinc oxide NPs. The films were also prepared by solution casting and characterized in terms of physicochemical properties, mechanical performance and barrier properties. The results showed that the incorporation of the nanorods into the films (3.5 w/w) leads to full absorption of UV light, while the tensile strength increased from 14 to 18 MPa and the oxygen permeability decreased from 151.03 to 91.52 $\mu m\ m^{-2}$ day^{-1}.

The simultaneous combination of starch with ZnO NPs and other nanostructures, as for example Ag NPs, has also been considered as a strategy to produce active films for packaging applications [131], taking advantage of the functionalities of both nanostructures.

Starch/SiO_2 based materials have been poorly explored in the last 5 years. Liu et al. [132] studied the effect of the addition of silica and a silane modified silica NPs on the morphology, structure and properties of a thermoplastic starch/poly/vinyl alcohol) blend concluding that the thermal stability and mechanical properties of the resulting hybrids increased with the increasing amount of modified silica. Following a different approach, starch-based hybrids with improved H_2O/O_2 selectivity for food packaging were prepared by solvent casting of hydroxypropyl starch/SiO_2 NPs mixtures [133]. More recently, Chotikhun et al. [134] evaluated the termite resistance of particle board panels composed of Eastern redcedar using SiO_2 NPs modified starch as binder. An average weight loss of 9.25% was observed for panels with 3% of NPs against 14.08% of the control sample, showing an increased level of resistance against the damage provoked termites.

Starch/iron oxide NPs have been far more investigated as magnetic nano-sorbents for different pollutants [136, 137, 141], nanosystems for theranostic applications [135, 142] and drug delivery systems [139, 143]. As an illustrative example, superparamagnetic starch-functionalized magnetic NPs (6–14 nm) obtained by a co-precipitation methodology and used for the removal of Cr (VI) from aqueous medium have been recently reported [136]. Under optimum pH conditions (pH = 2) the maximum adsorption capacity of the NPs was found to be 26.6 mg g^{-1}. Moreover, the NPs could be easily regenerated and reused because of their low adsorption activation energy. In a different study, Fernandes et al. [141] described novel magnetic nanosorbents composed of magnetite cores functionalized with starch and carrageenan hybrid siliceous shells for the uptake of paraquat from aqueous systems. These hybrid NPs showed superior paraquat removal, with a maximum value of 257.7 mg g^{-1}, placing it among the best systems for the removal of this herbicide from water. In a

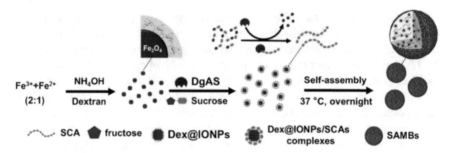

Fig. 3.3 Production of superparamagnetic amylose microbeads. Reprinted with permission from [144]. Copyright 2017 Elsevier

total different vein, Poorgholy et al. [142] developed stimuli-responsive starch/Fe_3O_4 based nanosystems for theranostic applications. For this purpose, starch was modified with itaconic anhydride, adsorbed onto the surface of iron oxide NPs, and finally co-polymerized with N-isopropylacrylamide. These nanosystems displayed temperature and magnetic responsiveness and a loading capacity of the nanosystems of about 74% towards the chemotherapeutic drug methotrexate. In another recent study, thiolated coated iron oxide NPS containing curcumin were produced and their cytotoxicity to lymphocytes and cancer cell lines was investigated. NPs with different concentrations of thiolated starch (1–5%) were produced and those with 5% loading showed a curcumin encapsulation efficiency of 78%. It was also concluded that the drug release was pH dependent and increased with the polymer concentration. The cytotoxicity tests revealed an excellent compatibility with lymphocyte cells but a considerable toxicity for the MFC7 and HepG2 cells.

Regarding the production of starch/iron oxide hybrids or nanocomposites, we can cite for example, the work developed by Gonçalves et al. [140] on the design of sophisticated magnetic polymer scaffolds for tendon regeneration by incorporating iron oxide NPs into a 3D structure of aligned starch/polycaprolactone (PCL) fibres fabricated by prototyping technology. The effect of an externally applied magnetic field was investigated and revealed that adipose stem cells undergo tenogenic differentiation, producing a Tenascin C and Collagen type I enriched matrix under magnetic stimulation. These scaffolds also evidenced good biocompatibility and integration within the surrounding tissues when implanted in ectopic rat models. Another interesting example refers to the production of starch-based magnetic polymers beads through co-precipitation of short chain amylose and dextran coated iron oxide NPs (Fig. 3.3) [144]. The obtained microbeads were readily functionalized with an antibody using a linker protein which showed great capture efficiency and specific target bacteria present in complicated food matrix. These beads exhibited also excellent biocompatibility and high magnetic sensitivity making them suitable for applications as, for example, immunomagnetic separation of target bacteria, bio-sensing and magnet-mediated delivery of drugs.

The available literature in starch/copper oxides hybrids is very scarce. Sun et al. [145] described a starch-assisted synthesis of cuprous oxide microcrystals for photo-catalytic applications. Octahedral Cu_2O structures with average size 1.6 μm showed the highest performance for the photodegradation of methyl orange with a pho-todegradation efficiency of 90.0% after visible light irradiation for 3.5 s. In another study, the green synthesis of CuO NPs using starch extracted from *Solanum tubero-sum* was investigated [146]. The CuO NPs were highly stable and spherical with a mean size of about 54 nm and showed considerable activity against several Gram-positive and negative bacterial strains. In addition, the MTT (3-(4,5-dimethylthiazol-2-yl)-2,5-diphenyltetrazolium bromide) assay showed anticancer activity for the syn-thesized CuO NPs against the MFC-7 cell line. In a totally different study, Ash-jari et al. [147] described a new method to produce starch-based polyurethane/CuO nanocomposite foams for bacterial infection control. The foam obtained at the opti-mal condition (starch dosage of 6 g *per* 100 g polyol and CuO NPs dosage of 1 g *per* 100 g polyol) had an open cell structure with high tensile strength (172.1 kPa) and efficient antimicrobial activity against nosocomial infections.

References

1. Muthiah M, Park I-K, Cho C-S. Surface modification of iron oxide nanoparticles by biocom-patible polymers for tissue imaging and targeting. Biotechnol Adv. 2013;31:1224–36.
2. Gautier J, Allard-Vannier E, Munnier E, Soucé M, Chourpa I. Recent advances in theranostic nanocarriers of doxorubicin based on iron oxide and gold nanoparticles. J Control Release. 2013;169:48–61.
3. Lima-Tenório MK, Gómez Pineda EA, Ahmad NM, Fessi H, Elaissari A. Magnetic nanopar-ticles: in vivo cancer diagnosis and therapy. Int J Pharm. 2015;493:313–27.
4. Yu X, Marks TJ, Facchetti A. Metal oxides for optoelectronic applications. Nat Mater. 2016;15:383–96.
5. Halbus AF, Horozov TS, Paunov VN. Colloid particle formulations for antimicrobial appli-cations. Adv Colloid Interface Sci. 2017;249:134–48.
6. Li S-S, Chen C-W. Polymer–metal-oxide hybrid solar cells. J Mater Chem A. 2013;1:10574.
7. Anasori B, Beidaghi M, Gogotsi Y. Graphene—transition metal oxide hybrid materials. Mater Today. 2014;17:253–4.
8. Boury B, Plumejeau S. Metal oxides and polysaccharides: an efficient hybrid association for materials chemistry. Green Chem. 2015;17:72–88.
9. Salama A. Polysaccharides/silica hybrid materials: new perspectives for sustainable raw mate-rials. J Carbohydr Chem. 2016;35:131–49.
10. Chauhan I, Aggrawal S, Chandravati C, Mohanty P. Metal oxide nanostructures incorporated/immobilized paper matrices and their applications: a review. RSC Adv. 2015;5:83036–55.
11. Ortelli S, Blosi M, Albonetti S, Vaccari A, Dondi M, Costa AL. TiO_2 based nano-photocatalysis immobilized on cellulose substrates. J Photochem Photobiol A Chem. 2013;276:58–64.
12. Li Y, Cao L, Li L, Yang C. In situ growing directional spindle TiO_2 nanocrystals on cellulose fibers for enhanced Pb^{2+} adsorption from water. J Hazard Mater. 2015;289:140–8.
13. El-Wakil NA, Hassan EA, Abou-Zeid RE, Dufresne A. Development of wheat gluten/nanocellulose/titanium dioxide nanocomposites for active food packaging. Carbohydr Polym. 2015;124:337–46.

14. Janpetch N, Vanichvattanadecha C, Rujiravanit C. Photocatalytic disinfection of water by bacterial cellulose/N–F co-doped TiO_2 under fluorescent light. Cellulose. 2015;22:3321–35.
15. Dal'Acqua N, de Mattos AB, Krindges I, Pereira MB, Barud HS, Ribeiro SJL, Duarte GCS, Radtke C, Almeida LC, Giovanela M, Crespo JS, Machado G. Characterization and application of nanostructured films containing Au and TiO_2 nanoparticles supported in bacterial cellulose. J Phys Chem C. 2015;119:340–9.
16. Li G, Nandgaonkar AG, Wang Q, Zhang J, Krause WE, Wei Q, Lucia LA. Laccase-immobilized bacterial cellulose/TiO_2 functionalized composite membranes: evaluation for photo-and bio-catalytic dye degradation. J Membr Sci. 2016;525:89–98.
17. Pang Z, Yang Z, Chen Y, Zhang J, Wang Q, Huang F, Wei Q. A room temperature ammonia gas sensor based on cellulose/TiO_2/PANI composite nanofibers. Colloids Surf A. 2016;494:248–55.
18. Fujiwara K, Kuwahara Y, Sumida Y, Yamashita H. Fabrication of photocatalytic paper using TiO_2 nanoparticles confined in hollow silica capsules. Langmuir. 2017;33:288–95.
19. Shandilya N, Capron I. Safer-by-design hybrid nanostructures: an alternative to conventional titanium dioxide UV filters in skin care products. RSC Adv. 2017;7:20430–9.
20. Khalid A, Ullah H, Ul-Islam M, Khan R, Khan S, Ahmad F, Khan T, Wahid F. Bacterial cellulose–TiO_2 nanocomposites promote healing and tissue regeneration in burn mice model. RSC Adv. 2017;7:47662–8.
21. Dumitriu C, Voicu SI, Muhulet A, Nechifor G, Popescu S, Ungureanu C, Carja A, Miculescu F, Trusca R, Pirvu C. Production and characterization of cellulose acetate—titanium dioxide nanotubes membrane fraxiparinized through polydopamine for clinical applications. Carbohydr Polym. 2018;181:215–23.
22. Martins NCT, Freire CSR, Neto CP, Silvestre AJD, Causio J, Baldi G, Sadocco P, Trindade T. Antibacterial paper based on composite coatings of nanofibrillated cellulose and ZnO. Colloids Surf A. 2013;417:111–9.
23. Khatri V, Halász K, Trandafiloví LV, Dimitrijevíc-Brankoví S, Mohanty P, Djokoví V, Csóka L. ZnO-modified cellulose fiber sheets for antibody immobilization. Carbohydr Polym. 2014;109:139–47.
24. Wang P, Zhao J, Xuan R, Wang Y, Zou C, Zhang Z, Wan Y, Xu Y. Flexible and monolithic zinc oxide bionanocomposite foams by a bacterial cellulose mediated approach for antibacterial applications. Dalton Trans. 2014;43:6762–8.
25. Gimenez AJ, Luna-Barcenas G, Sanchez IC, Yanez-Limon JM. Paper-based ZnO oxygen sensor. IEEE Sens J. 2015;15:1246–51.
26. Mun S, Maniruzzaman M, Ko H-U, Kafy A, Kim J. Preparation and characterisation of cellulose ZnO hybrid film by blending method and its glucose biosensor application. Mater Technol. 2015;30:150–4.
27. Fu F, Li L, Liu L, Cai J, Zhang Y, Zhou J, Zhang L. Construction of cellulose based ZnO nanocomposite films with antibacterial properties through one-step coagulation. ACS Appl Mater Interfaces. 2015;7:2597–606.
28. Grüneberger F, Künniger T, Huch A, Zimmermann T, Arnold M. Nanofibrillated cellulose in wood coatings: dispersion and stabilization of ZnO as UV absorber. Prog Org Coatings. 2015;87:112–21.
29. Nath BK, Chaliha C, Kalita E, Kalita MC. Synthesis and characterization of ZnO:CeO 2:nanocellulose:PANI bionanocomposite. A bimodal agent for arsenic adsorption and antibacterial action. Carbohydr Polym. 2016;148:397–405.
30. Zhao S-W, Zheng M, Zou X-H, Guo Y, Pan Q-J. Self-assembly of hierarchically structured cellulose@ZnO composite in solid–liquid homogeneous phase: synthesis, dft calculations, and enhanced antibacterial activities. ACS Sustain Chem Eng. 2017;5:6585–96.
31. Khalid A, Khan R, Ul-Islam M, Khan T, Wahid F. Bacterial cellulose-zinc oxide nanocomposites as a novel dressing system for burn wounds. Carbohydr Polym. 2017;164:214–21.
32. Wang D-C, Yu H-Y, Song M-L, Yang R-T, Yao J-M. Superfast adsorption–disinfection cryogels decorated with cellulose nanocrystal/zinc oxide nanorod clusters for water-purifying microdevices. ACS Sustain Chem Eng. 2017;5:6776–85.

33. Xu M, Wang H, Wang G, Zhang L, Liu G, Zeng Z, Ren T, Zhao W, Wu X, Xue Q. Study of synergistic effect of cellulose on the enhancement of photocatalytic activity of ZnO. J Mater Sci. 2017;52:8472–84.
34. Ibrahim NA, Eid BM, El-Aziz EA, Abou Elmaaty TM, Ramadan SM. Multifunctional cellulose-containing fabrics using modified finishing formulations. RSC Adv. 2017;7:33219–30.
35. Rajeswari A, Jackcina E, Christy S, Pius A. New insight of hybrid membrane to degrade Congo red and reactive yellow under sunlight. J Photochem Photobiol B Biol. 2018;179:7–17.
36. Jiang Y, Wang W, Li X, Wang X, Zhou J, Mu X. Enzyme-mimetic catalyst-modified nanoporous SiO_2–cellulose hybrid composites with high specific surface area for rapid H_2O_2 detection. ACS Appl Mater Interfaces. 2013;5:1913–6.
37. Song H, Zheng L. Nanocomposite films based on cellulose reinforced with nano-SiO_2: microstructure, hydrophilicity, thermal stability, and mechanical properties. Cellulose. 2013;20:1737–46.
38. Kim J-H, Kim J-H, Choi E-S, Yu HK, Kim JH, Wu Q, Chun S-J, Lee S-Y, Lee S-Y. Colloidal silica nanoparticle-assisted structural control of cellulose nanofiber paper separators for lithium-ion batteries. J Power Sources. 2013;242:533–40.
39. Evans E, Gabriel EFM, Benavidez TE, Coltro WKT, Garcia CD. Modification of microfluidic paper-based devices with silica nanoparticles. Analyst. 2014;139:5560–7.
40. He M, Duan B, Xu D, Zhang L. Moisture and solvent responsive cellulose/SiO_2 nanocomposite materials. Cellulose. 2015;22:553–63.
41. Fu J, Wang S, He C, Lu Z, Huang J, Chen Z. Facilitated fabrication of high strength silica aerogels using cellulose nanofibrils as scaffold. Carbohydr Polym. 2016;147:89–96.
42. Hakeem A, Zahid F, Duan R, Asif M, Zhang T, Zhang Z, Cheng Y, Lou X, Xia F. Cellulose conjugated FITC-labelled mesoporous silica nanoparticles: intracellular accumulation and stimuli responsive doxorubicin release. Nanoscale. 2016;8:5089–97.
43. Sheykhnazari S, Tabarsa T, Ashori A, Ghanbari A. Bacterial cellulose composites loaded with SiO_2 nanoparticles: Dynamic-mechanical and thermal properties. Int J Biol Macromol. 2016;93:672–7.
44. Cai Y, Hou X, Wang W, Liu M, Zhang J, Qiao H, Huang F, Wei Q. Effects of SiO_2 nanoparticles on structure and property of form-stable phase change materials made of cellulose acetate phase inversion membrane absorbed with capric-myristic-stearic acid ternary eutectic mixture. Thermochim Acta. 2017;653:49–58.
45. Nechyporchuk O, Bordes R, Köhnke T. Wet spinning of flame-retardant cellulosic fibers supported by interfacial complexation of cellulose nanofibrils with silica nanoparticles. ACS Appl Mater Interfaces. 2017;9:39069–77.
46. Garusinghe UM, Varanasi S, Garnier G, Batchelor W. Strong cellulose nanofibre–nanosilica composites with controllable pore structure. Cellulose. 2017;24:2511–21.
47. Chu G, Qu D, Zussman E, Xu Y. Ice-assisted assembly of liquid crystalline cellulose nanocrystals for preparing anisotropic aerogels with ordered structures. Chem Mater. 2017;29:3980–8.
48. Albertini F, Ribeiro T, Alves S, Baleizão C, Farinha JPS. Boron-chelating membranes based in hybrid mesoporous silica nanoparticles for water purification. Mater Des. 2018;141:407–13.
49. Anirudhan TS, Rejeena SR. Selective adsorption of hemoglobin using polymer-grafted-magnetite nanocellulose composite. Carbohydr Polym. 2013;93:518–27.
50. Chen L, Berry RM, Tam KC. Synthesis of β-cyclodextrin-modified cellulose nanocrystals (CNCs)@Fe_3O_4 @SiO_2 superparamagnetic nanorods. ACS Sustain Chem Eng. 2014;2:951–8.
51. Li J, Wang F, Shi D, Zhang Y, Shao Z. Multifunctional biopolymer nanoparticles for drug delivery and protein immobilization. Ferroelectrics. 2015;486:156–67.
52. Gomes D, Silva D, Toma SH, Menegatti De Melo F, Vieira L, Carvalho C, Magalhães A, Sabadini E, Domingues A, Santos D, Araki K, Toma HE. Direct synthesis of magnetite nanoparticles from iron(II) carboxymethylcellulose and their performance as NMR contrast agents. J Magn Magn Mater. 2015;397:28–32.

53. Wan C, Li J. Facile synthesis of well-dispersed superparamagnetic γ-Fe_2O_3 nanoparticles encapsulated in three-dimensional architectures of cellulose aerogels and their applications for Cr(VI) removal from contaminated water. ACS Sustain Chem Eng. 2015;3:2142–52.
54. Patra S, Roy E, Karfa P, Kumar S, Madhuri R, Sharma PK. Dual-responsive polymer coated superparamagnetic nanoparticle for targeted drug delivery and hyperthermia treatment. ACS Appl Mater Interfaces. 2015;7:9235–46.
55. Darwish MSA, Nguyen NHA, Ševců A, Stibor I, Smoukov SK. Dual-modality self-heating and antibacterial polymer-coated nanoparticles for magnetic hyperthermia. Mater Sci Eng, C. 2016;63:88–95.
56. Luo X, Lei X, Cai N, Xie X, Xue Y, Yu F. Removal of heavy metal ions from water by magnetic cellulose-based beads with embedded chemically modified magnetite nanoparticles and activated carbon. ACS Sustain Chem Eng. 2016;4:3960–9.
57. Sadasivuni KK, Ponnamma D, Ko H-U, Kim HC, Zhai L, Kim J. Flexible NO_2 sensors from renewable cellulose nanocrystals/iron oxide composites. Sens Actuators B Chem. 2016;233:633–8.
58. Bekaroğlu MG, İşçi Y, İşçi S. Colloidal properties and *in vitro* evaluation of Hydroxy ethyl cellulose coated iron oxide particles for targeted drug delivery. Mater Sci Eng, C. 2017;78:847–53.
59. Guo J, Filpponen I, Johansson L-S, Mohammadi P, Latikka M, Linder MB, Ras RHA, Rojas OJ. Complexes of magnetic nanoparticles with cellulose nanocrystals as regenerable, highly efficient, and selective platform for protein separation. Biomacromolecules. 2017;18:898–905.
60. Ortega GA, Pérez-Rodríguez S, Reguera E. Magnetic paper—based ELISA for IgM-dengue detection. RSC Adv. 2017;7:4921–32.
61. Zhou Z, Lu C, Wu X, Zhang X. Cellulose nanocrystals as a novel support for CuO nanoparticles catalysts: facile synthesis and their application to 4-nitrophenol reduction. RSC Adv. 2013;3:26066.
62. Hebeish A, Sharaf S. Novel nanocomposite hydrogel for wound dressing and other medical applications. RSC Adv. 2015;5:103036–46.
63. Booshehri AY, Wang R, Xu R. Simple method of deposition of CuO nanoparticles on a cellulose paper and its antibacterial activity. Chem Eng J. 2015;262:999–1008.
64. Sedighi Majid Montazer A. Tunable shaped N-doped CuO nanoparticles on cotton fabric through processing conditions: synthesis, antibacterial behavior and mechanical properties. Cellulose. 2016;23:2229–43.
65. Shankar S, Wang L-F, Rhim J-W. Preparation and properties of carbohydrate-based composite films incorporated with CuO nanoparticles. Carbohydr Polym. 2017;169:264–71.
66. Peng S, Fan L, Rao W, Bai Z, Xu W, Xu J. Bacterial cellulose membranes coated by polypyrrole/copper oxide as flexible supercapacitor electrodes. J Mater Sci. 2017;52:1930–42.
67. Almasi H, Jafarzadeh P, Mehryar L. Fabrication of novel nanohybrids by impregnation of CuO nanoparticles into bacterial cellulose and chitosan nanofibers: Characterization, antimicrobial and release properties. Carbohydr Polym. 2018;186:273–81.
68. Le Cunff J, Tomašić V, Wittine O. Photocatalytic degradation of the herbicide terbuthylazine: preparation, characterization and photoactivity of the immobilized thin layer of TiO2/chitosan. J Photochem Photobiol A Chem. 2015;309:22–9.
69. Xiao G, Su H, Tan T. Synthesis of core–shell bioaffinity chitosan–TiO_2 composite and its environmental applications. J Hazard Mater. 2015;283:888–96.
70. Kamari Y, Ghiaci M. Preparation and characterization of ibuprofen/modified chitosan/TiO_2 hybrid composite as a controlled drug-delivery system. Microporous Mesoporous Mater. 2016;234:361–9.
71. Tang Y, Hu X, Zhang X, Guo D, Zhang J, Kong F. Chitosan/titanium dioxide nanocomposite coatings: rheological behavior and surface application to cellulosic paper. Carbohydr Polym. 2016;151:752–9.
72. Yazdani MR, Bhatnagar A, Vahala R. Synthesis, characterization and exploitation of nano-TiO_2/feldspar-embedded chitosan beads towards UV-assisted adsorptive abatement of aqueous arsenic (As). Chem Eng J. 2017;316:370–82.

73. Shamsabadi AA, Seidi F, Salehi E, Nozari M, Rahimpour A, Soroush M. Efficient CO_2-removal using novel mixed-matrix membranes with modified TiO_2 nanoparticles. J Mater Chem A. 2017;5:4011–25.

74. Xu W, Xie W, Huang X, Chen X, Huang N, Wang X, Liu J. The graphene oxide and chitosan biopolymer loads TiO_2 for antibacterial and preservative research. Food Chem. 2017;221:267–77.

75. Zhang X, Xiao G, Wang Y, Zhao Y, Su H, Tan T. Preparation of chitosan-TiO_2 composite film with efficient antimicrobial activities under visible light for food packaging applications. Carbohydr Polym. 2017;169:101–7.

76. AL-Mokaram A, Yahya R, Abdi M, Mahmud H. The development of non-enzymatic glucose biosensors based on electrochemically prepared polypyrrole–chitosan–titanium dioxide nanocomposite films. Nanomaterials 2017;7:129.

77. Natarajan S, Lakshmi DS, Bhuvaneshwari M, Iswarya V, Mrudula P, Chandrasekaran N, Mukherjee A. Antifouling activities of pristine and nanocomposite chitosan/TiO_2/Ag films against freshwater algae. RSC Adv. 2017;7:27645–55.

78. Alizadeh B, Delnavaz M, Shakeri A. Removal of Cd(II) and phenol using novel cross-linked magnetic EDTA/chitosan/TiO_2 nanocomposite. Carbohydr Polym. 2018;181:675–83.

79. Samadi S, Moradkhani M, Beheshti H, Irani M, Aliabadi M. Fabrication of chitosan/poly(lactic acid)/graphene oxide/TiO_2 composite nanofibrous scaffolds for sustained delivery of doxorubicin and treatment of lung cancer. Int J Biol Macromol. 2018;110:416–24.

80. Cui H-F, Wu W-W, Li M-M, Song X, Lv Y, Zhang T-T. A highly stable acetylcholinesterase biosensor based on chitosan-TiO_2-graphene nanocomposites for detection of organophosphate pesticides. Biosens Bioelectron. 2018;99:223–9.

81. Krishnaveni R, Thambidurai S. Industrial method of cotton fabric finishing with chitosan–ZnO composite for anti-bacterial and thermal stability. Ind Crops Prod. 2013;47:160–7.

82. Díez-Pascual AM, Díez-Vicente AL. Wound healing bionanocomposites based on castor oil polymeric films reinforced with chitosan-modified ZnO nanoparticles. Biomacromol. 2015;16:2631–44.

83. Farzana MH, Meenakshi S. Exploitation of zinc oxide impregnated chitosan beads for the photocatalytic decolorization of an azo dye. Int J Biol Macromol. 2015;72:900–10.

84. Wang Y, Wang S, Tao L, Min Q, Xiang J, Wang Q, Xie J, Yue Y, Wu S, Li X, Ding H. A disposable electrochemical sensor for simultaneous determination of norepinephrine and serotonin in rat cerebrospinal fluid based on MWNTs-ZnO/chitosan composites modified screen-printed electrode. Biosens Bioelectron. 2015;65:31–8.

85. Zhao H, Lv P, Huo D, Zhang C, Ding Y, Xu P, Hu Y. Doxorubicin loaded chitosan–ZnO hybrid nanospheres combining cell imaging and cancer therapy. RSC Adv. 2015;5:60549–51.

86. Upadhyaya L, Singh J, Agarwal V, Pandey AC, Verma SP, Das P, Tewari RP. Efficient water soluble nanostructured ZnO grafted O-carboxymethyl chitosan/curcumin-nanocomposite for cancer therapy. Process Biochem. 2015;50:678–88.

87. Lu Z, Gao J, He Q, Wu J, Liang D, Yang H, Chen R. Enhanced antibacterial and wound healing activities of microporous chitosan-Ag/ZnO composite dressing. Carbohydr Polym. 2017;156:460–9.

88. Youssef AM, El-Sayed SM, El-Sayed HS, Salama HH, Dufresne A. Enhancement of Egyptian soft white cheese shelf life using a novel chitosan/carboxymethyl cellulose/zinc oxide bionanocomposite film. Carbohydr Polym. 2016;151:9–19.

89. Noshirvani N, Ghanbarzadeh B, Rezaei Mokarram R, Hashemi M. Novel active packaging based on carboxymethyl cellulose-chitosan-ZnO NPs nanocomposite for increasing the shelf life of bread. Food Packag Shelf Life. 2017;11:106–14.

90. Zhao S, You B, Jiang L. Oriented assembly of zinc oxide mesocrystal in chitosan and applications for glucose biosensors. Cryst Growth Des. 2016;16:3359–65.

91. Al-Naamani L, Dobretsov S, Dutta J, Burgess JG. Chitosan-zinc oxide nanocomposite coatings for the prevention of marine biofouling. Chemosphere. 2017;168:408–17.

92. Thirumalraj B, Rajkumar C, Chen S-M, Lin K-Y. Determination of 4-nitrophenol in water by use of a screen-printed carbon electrode modified with chitosan-crafted ZnO nanoneedles. J Colloid Interface Sci. 2017;499:83–92.

93. El-Mekawy RE, Jassas RS. Recent trends in smart and flexible three-dimensional cross-linked polymers: synthesis of chitosan–ZnO nanocomposite hydrogels for insulin drug delivery. MedChemComm. 2017;8:897–906.
94. Rasool K, Nasrallah GK, Younes N, Pandey RP, Abdul Rasheed P, Mahmoud KA. "Green" ZnO-interlinked chitosan nanoparticles for the efficient inhibition of sulfate-reducing bacteria in inject seawater. ACS Sustain Chem Eng. 2018;6:3896–906.
95. Elsayed A, Al-Remawi M, Maghrabi I, Hamaidi M, Jaber N. Development of insulin loaded mesoporous silica injectable particles layered by chitosan as a controlled release delivery system. Int J Pharm. 2014;461:448–58.
96. Feng W, Nie W, He C, Zhou X, Chen L, Qiu K, Wang W, Yin Z. Effect of pH-responsive alginate/chitosan multilayers coating on delivery efficiency, cellular uptake and biodistribution of mesoporous silica nanoparticles based nanocarriers. ACS Appl Mater Interfaces. 2014;6:8447–60.
97. Zhao H, Xu J-H, Wang T, Luo G-S. A novel microfluidic approach for preparing chitosan–silica core–shell hybrid microspheres with controlled structures and their catalytic performance. Lab Chip. 2014;14:1901–6.
98. Tian R, Qu Y, Zheng X. Amplified fluorescence quenching of lucigenin self-assembled inside silica/chitosan nanoparticles by Cl$^-$. Anal Chem. 2014;86:9114–21.
99. Qu Y, Han H, Zheng X, Guo Z, Li Y. Detection of surface pH of paper using a chitosan-modified silica fluorescent nanosensor. Sens Actuators B Chem. 2014;195:252–8.
100. Wang D, Romer F, Connell L, Walter C, Saiz E, Yue S, Lee PD, McPhail DS, Hanna JV, Jones JR. Highly flexible silica/chitosan hybrid scaffolds with oriented pores for tissue regeneration. J Mater Chem B. 2015;3:7560–76.
101. Tian R, Xian L, Li Y, Zheng X. Silica modified chitosan/polyethylenimine nanogel for improved stability and gene carrier ability. J Nanosci Nanotechnol. 2016;16:5426–31.
102. Roosen J, Van Roosendael S, Borra CR, Van Gerven T, Mullens S, Binnemans K. Recovery of scandium from leachates of Greek bauxite residue by adsorption on functionalized chitosan–silica hybrid materials. Green Chem. 2016;18:2005–13.
103. Zhao R, Li T, Zheng G, Jiang K, Fan L, Shao J. Simultaneous inhibition of growth and metastasis of hepatocellular carcinoma by co-delivery of ursolic acid and sorafenib using lactobionic acid modified and pH-sensitive chitosan-conjugated mesoporous silica nanocomplex. Biomaterials. 2017;143:1–16.
104. Pipattanawarothai A, Suksai C, Srisook K, Trakulsujaritchok T. Non-cytotoxic hybrid bioscaffolds of chitosan-silica: sol-gel synthesis, characterization and proposed application. Carbohydr Polym. 2017;178:190–9.
105. Rafigh SM, Heydarinasab A. Mesoporous chitosan–SiO_2 nanoparticles: Synthesis, characterization, and CO_2 adsorption capacity. ACS Sustain Chem Eng. 2017;5:10379–86.
106. Yu Z, Li B, Chu J, Zhang P. Silica in situ enhanced PVA/chitosan biodegradable films for food packages. Carbohydr Polym. 2018;184:214–20.
107. Blachnio M, Budnyak TM, Derylo-Marczewska A, Marczewski AW, Tertykh VA. Chitosan–silica hybrid composites for removal of sulfonated azo dyes from aqueous solutions. Langmuir. 2018;34:2258–73.
108. Ricardi NC, de Menezes EW, Valmir E, Benvenutti J, da Natividade Schöffer J, Hackenhaar CR, Hertz PF, Costa TMH. Highly stable novel silica/chitosan support for β-galactosidase immobilization for application in dairy technology. Food Chem. 2018;246:343–50.
109. Xiang X, Ding S, Suo H, Xu C, Gao Z, Hu Y. Fabrication of chitosan-mesoporous silica SBA-15 nanocomposites via functional ionic liquid as the bridging agent for PPL immobilization. Carbohydr Polym. 2018;182:245–53.
110. Prabhakar N, Thakur H, Bharti A, Kaur N. Chitosan-iron oxide nanocomposite based electrochemical aptasensor for determination of malathion. Anal Chim Acta. 2016;939:108–16.
111. Chaichi MJ, Ehsani M. A novel glucose sensor based on immobilization of glucose oxidase on the chitosan-coated Fe_3O_4 nanoparticles and the luminol–H_2O_2–gold nanoparticle chemiluminescence detection system. Sensors Actuators B Chem. 2016;223:713–22.

112. Chen G, Liu J, Qi Y, Yao J, Yan B. Biodiesel production using magnetic whole-cell biocatalysts by immobilization of *Pseudomonas mendocina* on Fe_3O_4-chitosan microspheres. Biochem Eng J. 2016;113:86–92.

113. Key J, Dhawan D, Cooper CL, Knapp DW, Kim K, Kwon IC, Choi K, Park K, Decuzzi P, Leary JF. Multicomponent, peptide-targeted glycol chitosan nanoparticles containing fer-rimagnetic iron oxide nanocubes for bladder cancer multimodal imaging. Int J Nanomed. 2016;11:4141–55.

114. Kostevsek N, Locatelli E, Garrovo C, Arena F, Monaco I, Nikolov IP, Sturm S, Zuzek Rozman K, Lorusso V, Giustetto P, Bardini P, Biffi S, Comes Franchini M. The one-step synthesis and surface functionalization of dumbbell-like gold–iron oxide nanoparticles: a chitosan-based nanotheranostic system. Chem Commun. 2016;52:378–81.

115. Lu J, Xu K, Yang J, Hao Y, Cheng F. Nano iron oxide impregnated in chitosan bead as a highly efficient sorbent for Cr(VI) removal from water. Carbohydr Polym. 2017;173:28–36.

116. Kong D, Wang N, Qiao N, Wang Q, Wang Z, Zhou Z, Ren Z. Facile preparation of ion-imprinted chitosan microspheres enwrapping Fe_3O_4 and graphene oxide by inverse sus-pension cross-linking for highly selective removal of copper(II). ACS Sustain Chem Eng. 2017;5:7401–9.

117. Zhang P, Fang X, Yan G, Gao M, Zhang X. Highly efficient enrichment of low-abundance intact proteins by core-shell structured Fe_3O_4-chitosan@graphene composites. Talanta. 2017;174:845–52.

118. Liu S, Kang M, Yan F, Peng D, Yang Y, He L, Wang M, Fang S, Zhang Z. Electrochemical DNA biosensor based on microspheres of cuprous oxide and nano-chitosan for Hg(II) detection. Electrochim Acta. 2015;160:64–73.

119. Senthil Kumar P, Selvakumar M, Babu SG, Jaganathan SK, Karuthapandian S, Chattopadhyay S. Novel CuO/chitosan nanocomposite thin film: facile hand-picking recoverable, efficient and reusable heterogeneous photocatalyst. RSC Adv. 2015;5:57493–501.

120. Cao C, Xiao L, Chen C, Cao Q. Magnetically separable Cu_2O/chitosan–Fe_3O_4 nanocompos-ites: Preparation, characterization and visible-light photocatalytic performance. Appl Surf Sci. 2015;333:110–8.

121. Chani MTS. Impedimetric sensing of temperature and humidity by using organic-inorganic nanocomposites composed of chitosan and a CuO-Fe_3O_4 nanopowder. Microchim Acta. 2017;184:2349–56.

122. Wahid F, Wang H-S, Lu Y-S, Zhong C, Chu L-Q. Preparation, characterization and antibacte-rial applications of carboxymethyl chitosan/CuO nanocomposite hydrogels. Int J Biol Macro-mol. 2017;101:690–5.

123. Nouri A, Yaraki MT, Ghorbanpour M, Agarwal S, Gupta VK. Enhanced antibacterial effect of chitosan film using montmorillonite/CuO nanocomposite. Int J Biol Macromol. 2018;109:1219–31.

124. Alzahrani E. Chitosan membrane embedded with ZnO/CuO nanocomposites for the pho-todegradation of fast green dye under artificial and solar irradiation. Anal Chem Insights. 2018;13:1–13.

125. Bao S-J, Lei C, Xu M-W, Cai C-J, Cheng C-J, Ming C. Li, Environmentally-friendly biomim-icking synthesis of TiO_2 nanomaterials using saccharides to tailor morphology, crystal phase and photocatalytic activity. CrystEngComm. 2013;15:4694–9.

126. Khodadadi B. Facile sol–gel synthesis of Nd, Ce-codoped TiO_2 nanoparticle using starch as a green modifier: structural, optical and photocatalytic behaviors. J Sol-Gel Sci Technol. 2016;80:793–801.

127. Wang X, Feng W, Wang W, Wang W, Zhao L, Li Y. Sodium carboxymethyl starch-based highly conductive gel electrolyte for quasi-solid-state quantum dot-sensitized solar cells. Res Chem Intermed. 2018;44:1161–72.

128. Baysal A, Kuznek C, Ozcan M. Starch coated titanium dioxide nanoparticles as a challenging sorbent to separate and preconcentrate some heavy metals using graphite furnace atomic absorption spectrometry. Int J Environ Anal Chem. 2018;98:45–55.

129. Kotharangannagari VK, Krishnan K. Biodegradable hybrid nanocomposites of starch/lysine and ZnO nanoparticles with shape memory properties. Mater Des. 2016;109:590–5.
130. Marvizadeh MM, Oladzadabbasabadi N, Mohammadi Nafchi A, Jokar M. Preparation and characterization of bionanocomposite film based on tapioca starch/bovine gelatin/nanorod zinc oxide. Int J Biol Macromol. 2017;99:1–7.
131. Kaur M, Kalia A, Thakur A. Effect of biodegradable chitosan-rice-starch nanocomposite films on post-harvest quality of stored peach fruit. Starch. 2017;69:1600208.
132. Liu Y, Mo X, Pang J, Yang F. Effects of silica on the morphology, structure, and properties of thermoplastic cassava starch/poly(vinyl alcohol) blends. J Appl Polym Sci 133 (2016).
133. Liu S, Li X, Chen L, Li L, Li B, Zhu J, Liang X. Investigating the H_2O/O_2 selective permeability from a view of multi-scale structure of starch/SiO_2 nanocomposites. Carbohydr Polym. 2017;173:143–9.
134. Chotikhun A, Hiziroglu S, Kard B, Konemann C, Buser M, Frazier S. Measurement of termite resistance of particleboard panels made from Eastern redcedar using nano particle added modified starch as binder. Measurement. 2018;120:169–74.
135. Zhang J, Shin MC, Yang VC. Magnetic targeting of novel heparinized iron oxide nanoparticles evaluated in a 9L-glioma mouse model. Pharm Res. 2014;31:579–92.
136. Singh PN, Tiwary D, Sinha I. Starch-functionalized magnetite nanoparticles for hexavalent chromium removal from aqueous solutions. Desalin Water Treat. 2016;57:12608–19.
137. Huo L, Zeng X, Su S, Bai L, Wang Y. Enhanced removal of As (V) from aqueous solution using modified hydrous ferric oxide nanoparticles. Sci Rep. 2017;7:40765.
138. Hamidian H, Tavakoli T. Preparation of a new Fe_3O_4/starch-g-polyester nanocomposite hydrogel and a study on swelling and drug delivery properties. Carbohydr Polym. 2016;144:140–8.
139. Saikia C, Das MK, Ramteke A, Maji TK. Effect of crosslinker on drug delivery properties of curcumin loaded starch coated iron oxide nanoparticles. Int J Biol Macromol. 2016;93:1121–32.
140. Gonçalves AI, Rodrigues MT, Carvalho PP, Bañobre-López M, Paz E, Freitas P, Gomes ME. Exploring the potential of starch/polycaprolactone aligned magnetic responsive scaffolds for tendon regeneration. Adv Healthc Mater. 2016;5:213–22.
141. Fernandes T, Soares SF, Trindade T, Daniel-da-Silva AL. Magnetic hybrid nanosorbents for the uptake of paraquat from water. Nanomaterials. 2017;7:68.
142. Poorgholy N, Massoumi B, Jaymand M. A novel starch-based stimuli-responsive nanosystem for theranostic applications. Int J Biol Macromol. 2017;97:654–61.
143. Saikia C, Das MK, Ramteke A, Maji TK. Controlled release of curcumin from thiolated starch-coated iron oxide magnetic nanoparticles: An *in vitro* evaluation. Int J Polym Mater Polym Biomater. 2017;66:349–58.
144. Luo K, Jeong K-B, Park C-S, Kim Y-R. Biosynthesis of superparamagnetic polymer microbeads via simple precipitation of enzymatically synthesized short-chain amylose. Carbohydr Polym. 2018;181:818–24.
145. Sun D, Du Y, Li Z, Chen Z, Zhu C, Liu S. Starch-assisted synthesis and photocatalytic activity of monosized cuprous oxide octahedron microcrystals. J Sol-Gel Sci Technol. 2016;78:347–52.
146. Alishah H, Pourseyedi S, Ebrahimipour SY, Mahani SE, Rafiei N. Green synthesis of starch-mediated CuO nanoparticles: preparation, characterization, antimicrobial activities and *in vitro* MTT assay against MCF-7 cell line. Rend Lincei. 2017;28:65–71.
147. Ashjari HR, Dorraji MSS, Fakhrzadeh V, Eslami H, Rasoulifard MH, Rastgouy-Houjaghan M, Gholizadeh P, Kafil HS. Starch-based polyurethane/CuO nanocomposite foam: antibacterial effects for infection control. Int J Biol Macromol. 2018;111:1076–82.
148. Yuan B, Zhang J, Mi Q, Yu J, Song R, Zhang J. Transparent cellulose–silica composite aerogels with excellent flame retardancy via an in situ sol–gel process. ACS Sustain Chem Eng. 2017;5:11117–23.
149. Wang Q, Wang Y, Chen L, Cai J, Zhang L. Facile construction of cellulose nanocomposite aerogel containing TiO_2 nanoparticles with high content and small size and their applications. Cellulose. 2017;24:2229–40.

150. Norfazilah W, Ismail W. Sol–gel technology for innovative fabric finishing—a review. J Sol-Gel Sci Technol. 2016;78:698–707.
151. Zhao S-W, Guo C-R, Hu Y-Z, Guo Y-R, Pan Q-J. The preparation and antibacterial activity of cellulose/ZnO composite: a review. Open Chem. 2018;16:9–20.
152. Sirvï JA, Visanko M, Heiskanen JP, Liimatainen H. UV-absorbing cellulose nanocrystals as functional reinforcing fillers in polymer nanocomposite films. J Mater Chem A Mater Energy Sustain. 2016;4:6368–75.
153. Wan C, Li J. Embedding ZnO nanorods into porous cellulose aerogels via a facile one-step low-temperature hydrothermal method. Mater Des. 2015;83:620–5.
154. Bagheri M, Rabieh S. Preparation and characterization of cellulose-ZnO nanocomposite based on ionic liquid ([C4mim]Cl). Cellulose. 2013;20:699–705.
155. Zhaochuang JM, Zhiguo S, Zhou WX. Preparation of ZnO–cellulose nanocomposites by different cellulose solution systems with a colloid mill. Cellulose. 2016;23:3703–15.
156. Fu F, Gu J, Cao J, Shen R, Liu H, Zhang Y, Liu X, Zhou J. Reduction of silver ions using an alkaline cellulose dope: straightforward access to Ag/ZnO decorated cellulose nanocomposite film with enhanced antibacterial activities. ACS Sustain Chem Eng. 2018;6:738–48.
157. Nypelö T, Rodriguez-Abreu C, Rivas J, Dickey MD, Rojas OJ. Magneto-responsive hybrid materials based on cellulose nanocrystals. Cellulose. 2014;21:2557–66.
158. Dhar P, Kumar A, Katiyar V. Magnetic cellulose nanocrystal based anisotropic polylactic acid nanocomposite films: influence on electrical, magnetic, thermal, and mechanical properties. ACS Appl Mater Interfaces. 2016;8:18393–409.
159. Marins JA, Soares BG, Barud HS, Ribeiro SJL. Flexible magnetic membranes based on bacterial cellulose and its evaluation as electromagnetic interference shielding material. Mater Sci Eng C Mater Biol Appl. 2013;33:3994–4001.
160. Zhang H, Luo X, Tang H, Zheng M, Huang F. A novel candidate for wound dressing: Transparent porous maghemite/ cellulose nanocomposite membranes with controlled release of doxorubicin from a simple approach. Mater Sci Eng, C. 2017;79:84–92.
161. Neves MC, Freire CSR, Costa BFO, Pascoal Neto C, Trindade T. Cellulose/iron oxide hybrids as multifunctional pigments in thermoplastic starch based materials. Cellulose. 2013;20:861–71.
162. Xiong R, Lu C, Wang Y, Zhou Z, Zhang X. Nanofibrillated cellulose as the support and reductant for the facile synthesis of Fe_3O_4/Ag nanocomposites with catalytic and antibacterial activity. J Mater Chem A. 2013;1:14910.
163. Jiao Y, Wan C, Bao W, Gao H, Liang D, Li J. Facile hydrothermal synthesis of Fe_3O_4@cellulose aerogel nanocomposite and its application in Fenton-like degradation of Rhodamine B. Carbohydr Polym. 2018;189:371–8.
164. Wysokowski M, Motylenko M, Stöcker H, Bazhenov VV, Langer E, Dobrowolska A, Czaczyk K, Galli R, Stelling AL, Behm T, Klapiszewski Ł, Ambro D, Nowacka M, Molodtsov SL, Abendroth SL, Meyer DC, Kurzydłowski KJ, Jesionowski T, Ehrlich H. An extreme biomimetic approach: hydrothermal synthesis of β-chitin/ZnO nanostructured composites. J Mater Chem A Mater Energy Sustain. 2013;1:6469–76.
165. Kumar PTS, Lakshmanan V-K, Raj M, Biswas R, Hiroshi T, Nair SV, Jayakumar R. Evaluation of wound healing potential of β-chitin hydrogel/nano zinc oxide composite bandage. Pharm Res. 2013;30:523–7.
166. Sahraee S, Ghanbarzadeh B, Milani JM, Hamishehkar H. Development of gelatin bio-nanocomposite films containing chitin and ZnO nanoparticles. Food Bioprocess Technol. 2017;10:1441–53.
167. Oun AA, Rhim J-W. Preparation of multifunctional chitin nanowhiskers/ZnO-Ag NPs and their effect on the properties of carboxymethyl cellulose-based nanocomposite film. Carbohydr Polym. 2017;169:467–79.
168. Bazhenov VV, Wysokowski M, Petrenko I, Stawski D, Sapozhnikov P, Born R, Stelling AL, Kaiser S, Jesionowski T. Preparation of monolithic silica–chitin composite under extreme biomimetic conditions. Int J Biol Macromol. 2015;76:33–8.

169. Smolyakov G, Pruvost S, Cardoso L, Alonso B, Belamie E, Duchet-Rumeau J. PeakForce QNM AFM study of chitin-silica hybrid films. Carbohydr Polym. 2017;166:139–45.
170. Ramos MLP, González JA, Albornoz SG, Pérez CJ, Villanueva ME, Giorgieri SA, Copello GJ. Chitin hydrogel reinforced with TiO_2 nanoparticles as an arsenic sorbent. Chem Eng J. 2016;285:581–7.
171. Wang Y, Pei Y, Xiong W, Liu T, Li J, Liu S, Li B. New photocatalyst based on graphene oxide/chitin for degradation of dyes under sunlight. Int J Biol Macromol. 2015;81:477–82.
172. Luo Y, Zhou Z, Yue T. Synthesis and characterization of nontoxic chitosan-coated Fe_3O_4 particles for patulin adsorption in a juice-pH simulation aqueous. Food Chem. 2017;221:317–23.
173. Abdellaziz LM, Hosny MM. Development and validation of spectrophotometric, atomic absorption and kinetic methods for determination of moxifloxacin hydrochloride. Anal Chem Insights. 2018;13:1–13.
174. Ahmad R, Mirza A. Facile one pot green synthesis of chitosan-iron oxide (CS-Fe_2O_3) nanocomposite: removal of Pb(II) and Cd(II) from synthetic and industrial wastewater. J Clean Prod. 2018;186:342–52.
175. Doğaç YI, Deveci İ, Teke M, Mercimek B. TiO_2 beads and TiO_2-chitosan beads for urease immobilization. Mater Sci Eng, C. 2014;42:429–35.
176. Zhang Y, Chen L, Liu C, Feng X, Wei L, Shao L. Self-assembly chitosan/gelatin composite coating on icariin-modified TiO_2 nanotubes for the regulation of osteoblast bioactivity. Mater Des. 2015;92:471–9.
177. Du Y, Li Y, Wu T. A superhydrophilic and underwater superoleophobic chitosan–TiO_2 composite membrane for fast oil-in-water emulsion separation. RSC Adv. 2017;7:41838–46.
178. Bui V, Park D, Lee Y-C. Chitosan combined with ZnO, TiO_2 and Ag nanoparticles for antimicrobial wound healing applications: a mini review of the research trends. Polymers. 2017;9:21.
179. Konwar A, Kalita S, Kotoky J, Chowdhury D. Chitosan–iron oxide coated graphene oxide nanocomposite hydrogel: a robust and soft antimicrobial biofilm. ACS Appl Mater Interfaces. 2016;8:20625–34.
180. Lin Y, Liu X, Xing Z, Geng Y, Wilson J, Wu D, Kong H. Preparation and characterization of magnetic Fe_3O_4–chitosan nanoparticles for cellulase immobilization. Cellulose. 2017;24:5541–50.
181. Soares PIP, Machado D, Laia C, Pereira LCJ, Coutinho JT, Ferreira IMM, Novo CMM, Borges JP. Thermal and magnetic properties of chitosan-iron oxide nanoparticles. Carbohydr Polym. 2016;149:382–90.
182. Shukla S, Jadaun A, Arora V, Sinha RK, Biyani N, Jain VK. *In vitro* toxicity assessment of chitosan oligosaccharide coated iron oxide nanoparticles. Toxicol Rep. 2015;2:27–39.
183. Goudarzi V, Shahabi-Ghahfarrokhi I, Babaei-Ghazvini A. Preparation of ecofriendly UV-protective food packaging material by starch/TiO_2 bio-nanocomposite: characterization. Int J Biol Macromol. 2017;95:306–13.

Chapter 4
Polysaccharides-Based Hybrids with Graphene

Graphene is a single layer of carbon atoms covalently bonded in a hexagonal crystalline structure, isolated for the first time by Geim and Novoselov from the University of Manchester (UK) [1]. Graphene is a flat-like single layer of hybridized sp^2 carbon atoms, which are densely packed onto each other into an ordered 2D honeycomb network (Fig. 4.1a), described by IUPAC as a single carbon layer of the graphite structure, describing its nature by analogy to a polycyclic aromatic hydrocarbon of quasi infinite size [2]. A unit hexagonal cell of graphene comprises two equivalent sub-lattices of carbon atoms, joined together by sigma (σ) bonds with a carbon-carbon bond length of 0.142 nm [3]. Each carbon atom in the lattice has a π-orbital that contributes to a delocalized network of electrons, making graphene sufficiently stable compared to other nanosystems and providing graphene with unique properties [4]. The applicability of graphene is based on the advantageous carbon network which provides this material with a combination of a large specific surface area, superior mechanical stiffness and flexibility, remarkable optical transmittance, high electronic and thermal conductivities, permeability to gases, as well as many other supreme properties [4].

However, the hydrophobic nature of graphene makes it incompatible with most biopolymers (and polysaccharides in particular) and tends to irreversibility agglomerate. For this reason, a graphene derivative, viz. graphene oxide (GO), is most often used to prepare polymer-based hybrids (also commonly refereed as nanocomposites). GO retains much of the properties of the highly valued pure graphene, but it is much easier and cheaper to prepare in bulk quantities, and its oxygen containing functionalities have demonstrated to be very attractive to grow chemical structures at its surface using diverse strategies [5]. The preparation of GO is mainly achieved through the chemical exfoliation of graphite given the low-cost and massive scalability of this method [6]. Since the structure of GO depends on the chemical oxidation process used to its preparation, the proposed GO structure models differ considerably as discussed in the critical review by Dreyer et al. [7]. Nevertheless, the structure of GO can be simplistically assumed as a graphene sheet functionalized mainly with hydroxyl (C–OH), carbonyl (C = O), carboxyl (COOH) and epoxide (C–O–C)

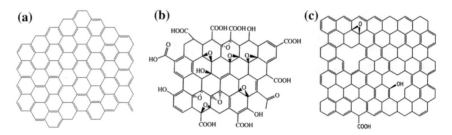

Fig. 4.1 Schematic representation of **a** graphene, **b** graphene oxide, and **c** reduced graphene oxide structures

moieties, among others [8] (Fig. 4.1b). The presence of these oxygen functionalities enhances the interfacial interaction between GO and polymer matrices, and even at very low amounts they improve considerably the physical properties of the resulting hybrids because of the improved stress transfer from the polymer matrix to GO [9]. Nevertheless, GO is frequently reduced (rGO) to produce a nanomaterial closer, in terms of structure, to pristine graphene (Fig. 4.1c). However, rGO owns several defects (holes) in the carbon network and some remaining oxygen functional groups. The degree of these defects depends mainly on the reduction method. Thus far, many reduction reagents have been used for this purpose (see for example Ref. [10]).

Frequently, the term graphene is used in a general and unprecise way creating some confusion and inconsistency. Besides the already referred structure lattice defects and different oxygen content (different C/O), also the number of carbon nanosheets layers can differ from 1 to 10, the average lateral dimension of the nanosheets can range from few nanometres to several micrometres. The need for unequivocal classification of these materials is well described by Wick et al. [11]. These authors present a concise discussion about the classification framework for graphene-based materials (GBMs) [11]. Consequently, in this chapter, the term GBMs is used generically when we are not referring to a specific carbon nanostructure.

Nanofillers such as GBMs are known to generate polymer-based hybrids with improved and often innovative mechanical but also thermal and electrical functional properties. The effects of the incorporation of GBMs in polymers have been extensively reviewed [12–14]. The final properties of the hybrids depend on numerous factors being the most important the interfacial bonding between the nanofiller and the matrix [15–17]. Interesting effects result from the addition of GBMs to polysaccharides, depending on the type of polysaccharide matrix and the interfacial interactions, which are directly related to their compatibility.

The present chapter summarizes the recent advances in the preparation of graphene-based hybrid materials with polysaccharides, their properties and applications, namely biomedical, water remediation, packaging and energy applications.

4.1 Biomedical Applications

The versatility polysaccharides to be used as biomaterials is attributable to their renewable nature, broad range of useful physical properties and facility to be shaped into different forms, such as capsules, beads, films, hydrogels and fibres [18], creating a multitude of uses, especially as drug delivery carriers and tissue engineering scaffolds. Despite these properties, their poor mechanical performance often limits their applications [15]. In this context, the combination of polysaccharides with GBMs allows to overcome not only the limited mechanical but also to improve specific physical properties and even biocompatibility [19].

Hydrogels are swollen, cross-linked networks that have great potential for use in biomedical applications. They are prepared by the cross-linked networks of hydrophilic polymers that swell in water and augment many times their original mass [20]. Several hydrogels have been developed based on polysaccharides towards regenerative medicine, drug delivery and tissue adhesives [21]. Hydrogels for biomedical applications are designed to resemble the characteristics of native extracellular matrix (ECM) and to provide three-dimensional (3D) supports for cellular growth and tissue formation. There are many polysaccharides used to form gel biomaterials being chitosan (CH), alginate (ALG), hyaluronan, cellulose and agarose, the most common [22–26]. However, hydrogels with improved mechanical properties are necessary during tissue regeneration to provide support and mechanical signalling to cells both in vitro and in vivo, for functional optimization of the new ECM being formed [27]. Thus, GBMs can be very good candidates to reinforce the hydrogels.

Cellulose hydrogels can be easily obtained by regeneration of dissolved cellulose. The search for environmentally friendly methods to dissolve cellulose led to the use of ionic liquids (ILs) [28, 29]. In this sequence, the preparation of cellulose/GBMs hydrogels have been pursuit using ILs [30]. Cellulose/rGO hydrogels may be prepared using a 1-butyl-3-methyl-imidazolium chloride solution to dissolve cellulose obtained from wood pulp and using deionized water as coagulant. Vitamin C was used as the reducing agent to convert GO into rGO directly in the IL. Mechanical strength and thermal stability of the resulting hydrogels were much higher in comparison with pure cellulose hydrogels, which is related with the doping ratio of rGO. Values of over four times higher compressive Young's modulus were obtained with 0.5 wt% doping with rGO [30].

Fan et al. [31] prepared a hydrogel through the cross-linking reaction between oxidized konjac glucomannan and carboxymethyl-CH with GO. The hydrogels showed rapid gelation, good swelling ability, appropriate water retention capacity and a stable 3D network structure, which all accomplish with the application of this hydrogel as wound dressings. The addition of 5 mg mL^{-1} of GO to the hydrogel increased the compressive strength and modulus by 144 and 296%, respectively, in relation to the hydrogel without GO. This behaviour is ascribed to the hydrogen bonding between GO sheets and the polysaccharides chains. Additionally, the hydrogels with GO have better biocompatibility.

ALG hydrogels functionalized with GO were explored by Marrella et al. [32] for the construction of cell laden hybrid materials. The cells used were human bone marrow mesenchymal stem cells (MSC). The results pointed out that the cell viability was statistically higher in hydrogels containing GO, while there was no significant difference between 0.5 and 2 wt% GO/ALG hydrogels up to 14 days. Furthermore, the mechanical properties of the hydrogel with 2 wt% GO were monitored up to one month under cell culture, demonstrating a significant improvement of the compressive elastic modulus reaching values of 300 kPa (6 times higher stiffness), which are close to those of articular tissues. The authors correlate this finding with the increased intermolecular hydrogen bonds over time between GO and ALG. The 3D GO/ALG hydrogels trigger cellular activity in vitro, as demonstrated by the statistically significant improvement of the viability of fibroblasts encapsulated in GO/ALG hydrogels and by the absence of cytotoxicity of suspended GO. GO/ALG hydrogels are therefore promising hybrids for articular tissue engineering, where biomechanical requirements are critical [32].

Nandgaonkar et al. [33] reported a one-pot in situ biosynthetic method to fabricate structurally controllable bacterial cellulose (BC) with rGO composites. GO (with concentrations between 0.05 and 0.5 mg mL^{-1}) was first reduced in an autoclave using yeast extract (a nutrient component for the cultivation of BC), then, the obtained rGO was dispersed in the BC culture medium, and a two-day old BC culture supernatant was added into the flask. Main cultivations were carried out at 30 °C for 7 days under stirring. As time progressed, the initial black colour solution became clearer indicating the incorporation of rGO nanosheets into BC membrane. The resulting BC/rGO composites allowed the preparation of different shapes: sealed structures in water, aerogels with a porous cross-section and aligned longitudinal structure and films. Atomic force microscopy (AFM) characterization showed that rGO incorporation into the BC culture media did not affect the BC nanofibrous structure being the rGO sheets randomly embedded in the BC fibrils. The green character of these hybrids foretells good results for several biomedical applications that include nerve cell signalling, muscle cell actuation, and heart cell innervation [33].

GBMs hold interest to integrate devices that can automatically respond to external stimulations such as electrical, chemical, photonic, thermal and other. The association of polysaccharides and GBMs is beneficial to build effective drug delivery systems (DDS) since it brings together biocompatibility and stimuli-responsive behaviour which can be crucial for DDS. In fact, Mura et al. [34] provided an interesting review article with recent advances in stimuli-responsive nanosystems, where the type of stimulus is distinguished between exogenous (variations in temperature, magnetic field, ultrasound intensity, light or electric pulses) or endogenous (changes in pH, enzyme concentration or redox gradients). Hereafter, some examples of DDS comprising natural polysaccharides and GBMs will be presented.

Deng et al. [35] designed a hybrid microcapsule (h-MC) using a layer-by-layer (LbL) methodology with three polysaccharides (ALG, CH and hyaluronan), together with iron oxide and GO, taking advantage of the electrostatic assembly of the different components. The h-MC were achieved using a sacrificial template of monodisperse spherical HCl-soluble melamine formaldehyde resin (MF) particles. Fe$_3$O$_4$-

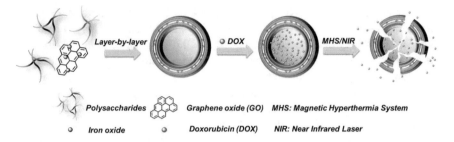

Fig. 4.2 Hybrid microcapsules (h-MC) via LbL assembly. Reprinted with permission from [35]. Copyright 2016 American Chemical Society

decorated GO (Fe$_3$O$_4$@GO) was synthesized by a co-precipitation method with a size of 191.2 ± 7.8 nm. Basically, the Fe$_3$O$_4$@GO nanosheet was incorporated within ALG/CH microcapsules with an hyaluronan shell. The Fe$_3$O$_4$@GO nanosheet is intended to be the functional layer between the alternating oppositely charged polysaccharides layers. The MF particles were etched by HCl to yield the hollow centre for the drug loading. Doxorubicin (DOX) was the selected anti-cancer drug, which was directly mixed with the microcapsules in PBS and incubated at 37 °C overnight. These capsules can be employed as a dual-responsive trigger induced by both NIR (near-infrared) and magnetic hyperthermia due to the presence of Fe$_3$O$_4$@GO nanosheets, while the hyaluronan ensure high bioavailability and tumour accumulation. The authors arrived at an interesting combination: by joining photothermal therapy (higher energy), magnetic hyperthermia (deeper penetration), and targeted chemotherapy (higher tumour accumulation) in a biocompatible carrier, generating a synergistic effect to efficiently annihilate cancer cells [35], as illustrated in Fig. 4.2.

The potential of GBM as drug nanocarriers is highlighted by Luo et al. [36] owing to their large specific surface area, as well as the chemical interactions between the drugs and the carbon network, namely by π–π stacking interactions, that prevent drugs from premature release outside the target cells. These authors studied the loading and release of ibuprofen (IBU) from BC/GO hybrid hydrogels. BC/GO hydrogels were prepared by adding GO (0.19 and 0.48 wt% with respect to BC) to the BC culture medium. The resulting hydrogels were immersed in an ethanol solution of IBU, followed by mild stirring at room temperature for 24 h. The drug-loaded samples were washed with deionized water and freeze dried. The in vitro release experiments were performed in PBS at pH 7.4 and simulated gastric fluid (SGF, pH 1.2) by dialysis bag technique at 37 °C. Results show that the drug can be simultaneously carried by both BC and GO, though IBU loading capacity increases with the GO content in BC/GO hydrogels. This was attributed to the increased specific surface area of BC/GO, as well as possible surface reaction between IBU and GO. The drug release studies showed a typical pH-sensitive release behaviour, that is faster in neutral and slower in acid media. In comparison with simple BC, BC/GO hydrogel demonstrated a more sustained released in both pH conditions [36].

Graphene Quantum Dots (GQDs) are fluorescent GBMs that retain the GO structural and physical properties but have smaller lateral dimensions (< 10 nm) and abundant periphery carboxylic groups [37]. GQDs have an outstanding ability in drug delivery and anti-cancer activity boost without any pre-modification due to their low toxicity, chemical inertness, solubility, functionality, crystallinity, two-dimensionality, biocompatibility, among other properties [37, 38]. Usually, traditional DDS could be visualized only by using organic fluorophores and semiconductor quantum dots to understand the cellular uptake, while in the case of GQDs, we can easily monitor movement in the cells in real time without employing external dyes considering its inherent fluorescence. The association of GQDs with hydrogels extend the DDS application, which can be used in different ways for oral, rectal, ocular, epidermal and subcutaneous applications. One of these examples is the use of carboxymethyl cellulose (CMC) to obtain flexible hybrid hydrogel films loaded with DOX, as proposed by Javanbakht et al. [39]. These hybrids were prepared through casting method of GQDs as a nanoparticle and CMC hydrogel as a polymeric matrix. The DOX loading in the CMC/GQD hydrogel films was made by swelling during 72 h in the dark with shaking. Results of swelling degradation, permeability and mechanical tests showed that, when compared with neat CMC, the CMC/GQDs hydrogel films have a great improvement in the tested properties. DOX loading and release studies revealed that the presence of GQDs in the CMC films caused a pH-sensitivity and consecutively prolonged the drug release. The hybrid without DOX showed non-significant toxicity against blood cancer cells (K562), meaning that the prepared DOX/CMC/GQD hybrids could possess a great potential as a long-lasting and high-efficiency anti-cancer agent [39].

Justin et al. [40] studied the association of GQDs with CH by creating biodegradable microneedles for electrically-stimulated and tracked transdermal DDS. As already mentioned, the advantage offered by GQDs is the ability to track, under fluorescent light, the diffusion of drugs in the body after delivery. In this study, the GQDs were bonded to lidocaine hydrochloride (LH) mainly through $\pi-\pi$ staking of the aromatic rings, with minor hydrogen bonding and subsequently integrated with CH by solution casting into moulds and dried to form microneedle arrays. Concentrations of GQDs of 0.25 to 2 wt% with respect to CH were tested, being the optimized concentration 1 wt%. These microneedle arrays tested resisted to a compressive force of at least 10 N, showing that they are strong enough to withstand the compressive force of insertion into human skin (i.e., 3.18 MPa or 1.6 N for the whole array of 100 microneedles [41]). The microneedle arrays have not broken or bent noticeably after the tests in comparison to their shape before the test. These arrays released the painkiller LH more substantially than pristine CH microneedles (68.3% of the available drug was released compared to 57.4%). Bovine serum albumin (BSA), as a model of a large molecular weight therapeutic agent, was also tested and the feasibility of realising BSA bonded to GQDs from the microneedle array assessed. The release was negligible under passive diffusion conditions, but was promoted under electrical stimulation, with an increase from 7.6 to 94.5% of the available drug after 24 h. These multifunctional materials provide a universal platform for stimulus-responsive and tracked delivery of both small and large molecular weight drugs [40].

Table 4.1 summarizes some representative studies dedicated to polysaccharides and GBMs hybrid materials developed for biomedical applications.

4.2 Water Remediation

In the last decades, nanotechnology has gained wide attention and several nanomaterials have been developed for water remediation, with GMBs being amongst the most promising. Indeed, these materials are gaining heightened attention as novel materials for environmental applications in general [42, 43]. The same can be stated for polysaccharides [44–47]. One of the drawbacks of many reported GBM aerogels is their brittle nature [43]. Also, most of them are formed by π–π staking assembly of GBMs nanosheets, which would greatly reduce the specific surface area and thus limit the applications [48]. In this context, the addition of GBM macrostructures to polysaccharides can be also designed to avoid these shortcomings.

One successful approach to overcome the previously reported drawbacks was achieved by adding previously electrospun cellulose acetate (CA) nanofibres to a GO aqueous suspension, which after a good homogenization was freeze dried [49]. The strong hydrogen bonding interactions between hydroxyl groups on the surface of CA nanofibres and the oxygen-containing groups on GO nanosheets allows obtaining aerogels with excellent mechanical performance. The prepared aerogels were tested for adsorption of dyes from contaminated waters showing high adsorption activity for cationic dyes due to strong electrostatic interactions with the aerogel. These aerogels were further modified for separation of oil-in-water emulsions. For this, polydopamine (PDA) and polyethylenimine (PEI) were added to the previous aerogels and compressed by 80% strain, which turned them into a robust membrane with pores of 5–10 μm due to the reshaping during compression, being suitable for separation of oil-in-water microemulsions. The contact angles in n-hexane and 1,2-dichloroethane on the surface are as high as 143.5 and 148.5°, respectively. The authors went even further and modified the initial aerogel with hexadecyltrimethoxysilane vapour yielding a hydrophobic aerogel. This latest aerogel presents a high capacity for the removal of oils floating on or sinking in water. Furthermore, it shows a remarkable compressive elasticity, being able to recover to the original shape after being compressed by 80% making possible the oils recovery. This work by Xiao et al. [49] gives an excellent example of the versatility of polysaccharides/GBMs aerogels, which can be prepared with different pore size and surface chemistry allowing to modulate their application.

Highly compressible cellulose nanofibrils (CNFs)/GO aerogels were prepared by bidirectional freeze drying allowing the preparation of ultra-light anisotropic aerogels [50]. Different CNFs/GO homogeneous mixtures were bidirectional freeze dried, followed by modification with n-dodecyltriethoxysilane vapour to prepare superhydrophobic aerogels. With respect to the average pore size and pore aspect ratio, both increased as the content of GO increased until 20 wt%. This composition was cho-

Table 4.1 Examples of different polysaccharides and GBMs hybrids for biomedical applications

Polysaccharide(s)	GBMs	Methodology	Main outcome/application	References
Cellulose	rGO	Regeneration of the mixture of wood pulp and rGO from ionic liquid using water as coagulant. Vitamin C was the reducing agent to prepare rGO directly in the IL. The hydrogels were freeze-dried	Biomedicine, environment and energy	[30]
Cellulose nanocrystals (CNCs)	rGO	Poly(lactic acid)/CNCs/rGO nanocomposites films were prepared through solution casting method	Antibacterial response and good biocompatibility	[86]
CMC	GQDs	Solvent casting using glycerol as a plasticizer and epichlorohydrin as a cross-linker of CMC and GQDs	DDS of doxorubicin for cancer cells	[39]
BC	GO	Hydroxyapatite (HAp) was prepared in the presence of GO (GOHAp) and dispersed in ethanol. Wet BC membrane was added to the above solution and freeze dried	Scaffold in bone defects	[20]
	GO	Sonochemical method using crushed BC membranes by high speed homogenizer	Biomedical applications (antibacterial)	[32]

(continued)

Table 4.1 (continued)

Polysaccharide(s)	GBMs	Methodology	Main out-come/application	References
	rGO	In situ synthesis of BC in the presence of rGO	Nerve cell signalling, muscle cell actuation, and heart cell innervation	[33]
	GO	BC/GO hybrids were prepared by adding a GO suspension to the BC culture medium	DDS of ibuprofen for Inflammatory diseases	[36]
CH	GO	Self-assembly and simultaneous reduction and cross-linking, in which GO serves as the cross-linking agent for CH	Bone tissue materials	[23]
	GO	The drug was loaded on GO–CH via phosphate esterification	DDS of dexamethasone phosphorylated for inflammations	[87]
	GQDs	GQDs were bonded to the drug and integrated with chitosan by solution casting	DDS of lidocaine hydrochloride for local anaesthetic, heart arrhythmia, epilepsy	[40]
CH and gelatin	GO	GO was sonicated with a CH/gelatin mixture in acetic acid. GO/CH/Gn scaffolds were obtained by freeze drying	Porous structure mimicking the natural architecture of bone	[83]
CH and hyaluronan	GO	GO/hyaluronan/CH films were prepared by LbL technique using the hand dipping method	Blood contacting materials (anti-thrombogenicity)	[82]

(continued)

Table 4.1 (continued)

Polysaccharide(s)	GBMs	Methodology	Main out-come/application	References
	GO	Solution mixing in aqueous acetic acid solution followed by dialysis and freeze drying	DDS of SNX-2112 (anti-cancer drug)	[88]
Carboxymethyl-CH	GO	Cross-linking reaction between carboxymethyl-CH and oxidized konjac glucomannan with GO as an additive	Increased compressive strength and modulus, better biocompatibility	[31]
ALG	GO	ALG hydrogels were synthesised with calcium chloride as cross-linking agent and 1 wt% of GO	Tissue engineering, bioprocess engineering, drug carriers	[84]
	GO	Cylindrical holes were made inside agarose gels and the GO/ALG mixtures were poured inside them. The mixtures were chemically cross-linked with calcium ions	Articular tissue	[32]
ALG, CH, hyaluronan	GO	Hybrid microcapsule produced by a LbL method	DDS of DOX for cancer cells	[35]
ALG, gellan and amylopectin	GO	The scaffolds were prepared by mixing all the components in aqueous solution and then freeze dried	Tissue engineering	[85]

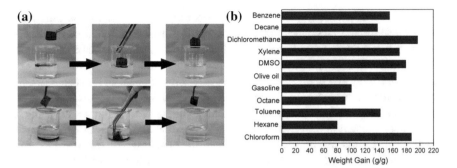

Fig. 4.3 a Demonstration of selective light oil (dyed hexane) absorption atop water and heavy oil (dyed chloroform) absorption underwater using cellulose/GO aerogel. **b** The absorption capacity of cellulose/GO aerogel towards various oils and chemical solvents. Reprinted with permission from [50]. Copyright 2018 Elsevier

sen to selectively absorb light oil (hexane) over water and heavy oil (chloroform) under water. As shown in Fig. 4.3a, the aerogel rapidly absorbed hexane from the water surface and remained floating. When used to absorb oil under water, an air layer was formed around the aerogel, protecting it from getting wet, which proves its high hydrophobicity. Oil was fast absorbed as soon as the two substances came into contact. The absorption capacity ranged from 80 to 197 times the aerogel weight, depending on the surface tension and density of the liquids tested. A final remark to mention the reusability of the aerogel, with less than 3% decrease in absorption capacity by extraction and a less than 10% decrease in absorption capacity by mechanical squeezing when reused for 10 cycles [50].

Aerogels composed of GO and microcrystalline cellulose (GO/MCC) were prepared by dissolving GO and lithium bromide in distilled water, followed by MCC dissolution in the previously prepared solution to obtain a gel, which was then converted to an aerogel through a solvent exchange with deionized water and freeze-drying treatment [51]. This solvent system has the advantage of promoting the gelation of cellulose in 1 min in the presence of GO, which consequently ensure the homogeneous dispersion of GO in the final hybrid aerogels. The GO/MCC aerogels exhibit a 3D porous structure that can be adjusted by varying the GO content. The adsorption ability of the aerogels was measured using methylene blue as a model adsorbate. Interestingly, the GO/MCC 0.3 wt% ratio presented an adsorption ability of 2630 mg g^{-1}, which is significantly higher when compared with the pure GO aerogel (824 mg g^{-1}). The MCC aerogel presents a negligible adsorption ability value less than 10 mg g^{-1} [51].

Yao et al. [52] prepared a CNF/GO aerogel via a one-step ultrasonication method with adsorption ability for the removal of 21 types of antibiotics from water. The as-prepared CNF/GO aerogel possesses interconnected 3D network microstructure, in which the 2D GO nanosheets were intimately linked along CNF through hydrogen bonds. The aerogel exhibited superior adsorption ability towards the antibiotics. The removal percentages of the antibiotics were higher than 69% and the sequence of six categories antibiotics according to the adsorption efficiency was as follows: tetracyclines > quinolones > sulfonamides > chloramphenicols > β-lactams > macrolides. The adsorption mechanism was proposed to be based on electrostatic attraction, p–π interaction, π–π interaction and hydrogen bonding. The authors also refer that the CNF/GO aerogel possesses reusability and can be easily removed from water presenting a great potential for future implementation [52].

Heavy metals are between the most common pollutants found in wastewater and can be accumulated in the environment and living tissues posing a threat to human health [53]. GBMs are promising candidates for heavy metals removal [43, 54]. Some authors are starting to study the formulation of polysaccharides with GBM also for this purpose, either in the form of hydrogels [55] or membranes [53, 56]. The association between GBMs and polysaccharides are starting to be explored for heavy metals removal from wastewater. The role of CH as nanofiller of GO to prepare materials presenting improved Hg (II) adsorption properties was explored by Kyzas et al. [57]. The oxygen functional groups of GO were combined and interacted with the amino groups of CH (or/and in CH with magnetic nanoparticles (Fe_3O_4) (GO/mCH)) creating new sites for Hg (II) adsorption. It is known that amino groups are responsible for metal ion binding through chelation mechanisms and may also contribute to adsorption process [57]. Three materials in powdered form (GO, GO/CH and GO/mCH) were studied and compared with respect to Hg (II) adsorption ability. At 25 °C, the adsorption capacity of the initial GO was 187 mg g^{-1}, which increased to 381 mg g^{-1} for GO/CH and to 397 mg g^{-1} for GO/mCH. The advantage of using GO/mCH is related with the easier separation from the aqueous medium. A similar composition of GO with CH and magnetic nanoparticles was successfully tested for Cr (IV) water removal [58].

Table 4.2 summarizes recent examples of hybrid materials composed of polysaccharides and GBMs for water remediation.

4.3 Packaging Applications

In food packaging applications, it is essential that protective films present good mechanical properties and reduced permeation to oxygen and other gases and volatile compounds. Furthermore, films should have good thermal stability, good transparency and anti-microbial activity to increase the shelf-life of food products. The GBMs incorporation in polysaccharides proved to have a positive effect in the mechanical and barrier properties of the resulting materials. For that, it is crucial to ensure a good dispersion of GBMs into the polysaccharide matrix, which is usually achieved at low GBMs loading, and guarantee good compatibility between the GBMs and host matrix, for good transfer load. Furthermore, the alignment and orientation of GBMs inside the polysaccharide matrix is also important. Several studies reported the preparation of cellulose [59], CH [60], ALG [61] and starch [62] hybrid films with GBMs by solvent casting.

Table 4.2 Overview of different polysaccharides and GO hybrids for water remediation

Polysaccharide	Methodology	Pollutant	References
Cellulose	Thermal treatment of the mixture in an autoclave and drying under vacuum followed by mixing with KOH and heating	Organophosphorus pesticide	[90]
	Cross-linking of cellulose and GO using epichlorohydrin	Metal ions, namely Cu^{2+}	[55]
	One-step ultrasonication and freeze-drying	21 kinds of antibiotics	[52]
	Cross-linking of cellulose and GO using ethylenediamine	Cu (II) and Pb (II)	[92]
α-cellulose	A mixture of α-cellulose, GO, NaOH and urea was vigorously stirred and converted to powder	Uranium (VI)	[91]
MCC	GO/cellulose membranes (pressed and non-pressed membranes)	Co (II), Ni (II), Cu (II), Zn (II), Cd (II)	[53]
CNFs	Bidirectional freeze drying and chemical vapour deposition to gain super-hydrophobicity	Selective oil absorption and recovery	[50]
CA	Freeze-drying	Cationic dyes Oil/water separation	[49]
CH	Cross-linking of GO and CH solutions using glutaraldehyde	Hg (II)	[57]
ALG	Freeze-drying and ionic cross-linking	Oil	[89]
	Freeze-drying and ionic cross-linking	Oil	[89]
	GO/ALG beads were formed by coagulation	Cu (II)	[93]

CH/GO films with improved barrier and thermo-mechanical properties were prepared by Ahmed et al. [63]. For 2 wt% of GO content, the value of the tensile strength increased around 119% and the water vapour permeability (WVP) was reduced in about 56%, while oxygen permeability (OP) decreased *ca.* 65% in comparison with pure CH films. The excellent dispersion of GO in the CH matrix was pointed as the main reason for the mechanical and barrier properties enhancements. The good compatibility between the GBMs and the host matrix enabled a good transfer load, which enhanced the mechanical properties. The alignment of nanofillers perpendicularly to the direction of diffusion maximized the pathway tortuosity and delayed the diffusion of gases. A ternary system of poly(vinylpyrrolidone) (PVP)/CH/GO was also prepared by casting method. The incorporation of 2 wt% of GO into the CH/PVP films with a 1:1 proportion promoted an enhancement of 130 and 109% in Young's modulus and tensile strength, respectively, in comparison to the CH/PVP films [64].

The effect of GBMs content on the mechanical and barrier properties of starch/GO and starch/rGO hybrid films was estimated by Ma et al. [65]. For that, GO and rGO were mixed with plasticized starch (with glycerol) at different loading levels (0, 0.5, 1, 2, 3 and 4 wt% for GO and 0, 2, 4, 6 and 8 wt% for rGO). Globally, both GO and rGO acted as reinforcement in starch matrices, until a certain level of incorporation (2 and 6 wt% for GO and rGO, respectively). The effectiveness of GO is higher compared to rGO due to the higher content of oxygen functionalities in GO, which contribute to a higher hydrogen bonding between GO and starch [65].

Table 4.3 summarizes the effect of GBMs on the mechanical and barrier properties (OP and WVP) of different polysaccharides. As can be observed a low amount of GBMs addition to polysaccharides films enhance the main required properties for good performance films for packaging applications.

4.4 Energy Applications

Polysaccharides/GBMs have started to be used for the development of new and high-performance energy devices. Polysaccharides can produce lightweight and flexible materials, while GBMs promote the electrochemical performance and strength [66]. In energy applications (*e.g.*, conductive papers, energy storage devices and supercapacitors), cellulose is the most studied polysaccharide. Herein, special focus will be placed on the specific capacitance and stability of charge-discharge cycles, since the main challenge in the production of components for energy storage devices relies in producing flexible and freestanding materials with high specific capacitance and long cycle lives [67, 68].

Cellulose/GBMs paper films are typically mentioned for the above referred applications [68, 69]. Cellulose nanopapers can be easily obtained via classical papermaking or similar processes (*e.g.* vacuum filtration) using cellulose obtained from CNF wood pulp and BC pellicles. The incorporation of GBMs into the cellulose matrix, either pre-mixed with cellulose fibres [70, 71] or depositing GBMs suspensions as a coating onto paper surface [72, 73], can improve electrochemical performance and

Table 4.3 Effect of GBMs on the mechanical and barrier properties of different polysaccharides

Polysaccharide	Filler	Mechanical properties improvement (%)		Barrier properties improvement (%)		References
		Tensile strength	Young's modulus	OP	WVP	
Cellulose	GO 4 wt%	56	30	71.6	–	[96]
CNFs	GO 0.6 wt%	43	16	73	–	[59]
Regenerated cellulose	GO 1.64 vol.%	67	68	–	–	[94]
CMC	GO 7 wt%	268	623	–	–	[95]
CH	GO 2 wt%	119	–	65	56	[63]
	GO 2 wt%	109	130	–	–	[64]
	GO 1 wt%	160	–	90	–	[60]
	GO 0.7 wt%	30	57	32	–	[97]
	GO 6 wt%	80	52	–	–	[98]
Starch	rGO 1 wt%	120	–	26	–	[99]
	GO 2 wt%	92.6	–	–	–	[65]
	Graphene 0.5 wt%	49	20	–	–	[100]
	GO 0.7 wt%	35	62	–	66	[101]
	Graphene 0.3 wt%	24.5	14	50	–	[102]
ALG	GO 6 wt%	592	392	–	–	[61]
	Tetraethylenepentamine-GO 1.0 wt%	68	–	–	–	[103]

strength. Kang et al. [72] prepared graphene/cellulose paper electrodes with 3.2 wt% of graphene obtaining a high specific capacitance of 252 F g^{-1} and good cycle charge-discharge stability, of up to 99% capacitance over 5000 cycles. The association of rGO to a metal hydroxide can enhance even more the performance of the final material. Ma et al. [74] prepared a ternary system of BC/rGO/Ni(OH)$_2$, where BC paper prepared by vacuum filtration of BC fibres was coated with as-prepared rGO-wrapped flowery Ni(OH)$_2$ obtained by a hydrothermal process. A flexible and freestanding Ni(OH)$_2$/rGO/BC film was obtained with specific capacitance of 877.1 F g^{-1} and excellent cycling stability (93.6% capacitance retention after 15,000 cycles).

The aerogels of polysaccharides/GBMs are highly porous materials with low densities and large specific areas which facilitate the access to the electrolyte solutions, and thus enhancing its electrochemical performance [69]. The role of GBMs is to increase the surface area and improve the electrochemical performance by increasing the electrical conductivity and mechanical stability. The behaviour of these nanocomposites is strongly dependent on the material design. For this application, the spatial orientation of the graphene sheets in cellulose/GBMs hybrids is a matter of great importance. It is documented that, for the case of graphene sheets, these can easily suffer agglomeration (because contrary to GO, graphene does not have oxygen functional groups), which can result in the reduction of surface area and compromise the electrochemical performance. To avoid agglomeration, the authors proposed the use of cellulose spacers and thus improve the electrochemical performance of the material [75–77].

The use of CNFs as nanospacer was also evaluated in a study conducted by Gao et al. [78]. The authors prepared CNFs/rGO hybrid aerogel by supercritical CO$_2$ drying and concluded that CNFs can effectively reduce the strong interactions between the graphene nanosheets maintaining, simultaneously, their intrinsic characteristics and consequently allowing good electrochemical performance of the material. The device capacitance still retained about 99.1% of the initial capacity (207 F g^{-1}) after 5000 charge–discharge cycles.

Two studies performed by Liu and co-workers [76, 77] reported the fabrication of a 3D structure by covalent intercalation of BC fibrils and GO sheets via one-step esterification between carboxylic groups of GO and hydroxyl groups from BC. In the first study, they noticed the importance of covalent bonding, by comparing the performance between BC/GO composite prepared by simple physical mixing and by covalent bonding. The results reported a highly improvement in the electrical conductivity conferred by the efficient conductive network formed with the covalent interpenetration between the GO and the BC nanofibrils which effectively avoided the aggregation of the GO sheets [76]. In the later study, the authors reported the development of a ternary system, Fig. 4.4, where a chemically bonded BC/GO hybrid composite was coated with a conductive polymer, polypyrrole, PPy [77]. Comparing the binary (BC/GO) and ternary (BC/GO/PPy) systems, an improvement of 248% in specific capacitance was achieved for the ternary system.

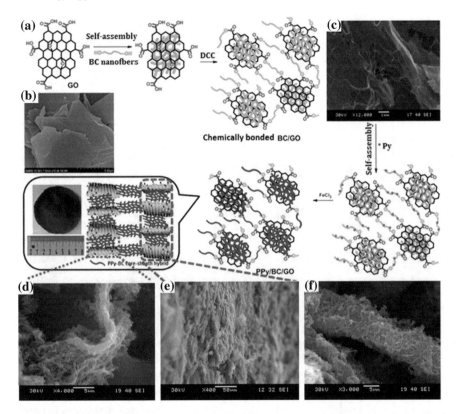

Fig. 4.4 **a** Self-assembly of BC nanofibres on GO surface, BC and GO cross-linking, reduction of GO, self-assembly of Py on BC nanofibre surface, and in situ polymerization to prepare PPy/BC/GO composites. **b** SEM image of pristine GO. The cross-sectional view of SEM image of **c** cross-linked BC/GO via the covalent intercalation of GO sheets with BC, **d** a single layer of PPy/BC/GO hybrid, **e** multilayers of PPy/BC/GO as stacking, and **f** PPy/BC core/sheath hybrid as linkage between stacking. Reprinted with permission from [77]. Copyright 2015 American Chemical Society

High performance materials were also obtained from ternary systems resulting from the combination of bacterial cellulose, polyaniline and graphene (BC/PANI/G) [79]. For the BC/PANI/G system, Liu et al. took advantage of the functional surface of BC to polymerize PANI. A layer of graphene was deposited by vacuum filtration on the surface of BC/PANI making a film with 477 F g^{-1} of specific capacitance and retaining 97% of that initial capacitance after 1000 cycles [79]. Although promising results were obtained, the non-uniform dispersion of graphene in the system and the lack of suitable pore size may have contributed to decrease the electrical conductivity and greatly contributed to the loss of mechanical integrity and stability. A different and improved in situ biosynthesis method of BC/graphene was presented by Luo et al. to preserve the 3D intrinsic network of BC [80]. For that, the authors added graphene suspensions to BC culture medium and found that the structure of graphene remained unchanged after the procedure. Furthermore, graphene sheets were uniformly

Table 4.4 Summary of the main properties of polysaccharides/GBMs hybrids for energy applications

Material composition	Preparation method	Main properties		References
		Specific capacitance	Cycling stability	
Cellulose/rGO	Papermaking process	212 F g^{-1}, current density of 1.0 A g^{-1}	94%—14,000 cycles	[70]
Cellulose/graphene paper	Vacuum filtration	80 mF cm^{-2}	95%—5000 cycles	[73]
Cellulose/graphene paper	Infiltration	252 F g^{-1}, current density of 1 A g^{-1}	> 99%—5000 cycles	[72]
Cellulose/graphene/polyacrylamide	Papermaking process		> 99%—1600 cycles current density of 1 A g^{-1}	[71]
CNFs/rGO	Supercritical CO$_2$	207 F g^{-1}	99.1%—5000 cycles	[78]
CNFs/GO/carbon nanotubes	Freeze-drying	252 F g^{-1}, current density of 0.5 A g^{-1}	> 99.5%—1000 cycles current density of 1 A g^{-1}	[104]
CNFs/graphene	Hydrothermal self-assembly, freeze-drying, carbonization	300 F g^{-1}, scan rate of 5 mV/s	95.4%—3000 cycles	[107]
CNFs/LiFePO$_4$/graphene	Vacuum filtration		95%—60 cycles	[105]
BC/polypyrrole/GO	In situ polymerization	492 F g^{-1}, current density of 1 A g^{-1}	93.5%—2000 cycles	[77]
BC/graphene/PANI	LbL method followed by in situ polymerization	645 F g^{-1}, current density of 1 A g^{-1}	82.2%—1000 cycles	[80]
BC/rGO	Impregnation of GO sheets into BC, freeze-drying Aannealing with N$_2$	216 F g^{-1}, current density of 1 A g^{-1}	86%—10,000 cycles	[106]
BC/GO/Ni(OH)$_2$	Hydrothermal treatment, vacuum filtration	877.1 F g^{-1}, current density of 5 mA cm^{-2}	93.6%—15,000 cycles	[74]

(continued)

Table 4.4 (continued)

Material composition	Preparation method	Main properties		References
		Specific capacitance	Cycling stability	
BC/GO	Impregnation of GO sheets into BC pellicle, freeze-drying, carbonization, annealing with N_2	216 F g^{-1}, current density of 1 A g^{-1}	86%—10,000 cycles	[106]
CH/GO	Hydrothermal treatment in the presence of copper chloride (ionic liquid)	356 F g^{-1}	80%—200,000 cycles	[81]
	Freeze drying of CH/GO hydrogel, carbonization in N_2 atmosphere	320 F g^{-1}, current density of 1 A g^{-1}	96%—2000 cycles	[108]
	Supercritical CO_2 carbonization in N_2 atmosphere	244.4 F g^{-1}, current density of 0.2 A g^{-1}	96.2%—5000 cycles	[109]
CH/GO-carbon nanotubes/PANI	In situ polymerization	609.2 F g^{-1}, scan rate of 10 mV s^{-1}	96%—500 cycles	[110]
PANI/dialdehyde starch-rGO	In situ oxidative polymerization	499 F g^{-1}, current density of 0.5 A g^{-1}	83%—1000 cycles	[111]
ALG/graphene nanosheets	Freezing, solvent exchange (ice-ethanol), ethanol drying	114.12 F g^{-1}, current density of 1 A g^{-1}	82%—1000 cycles	[112]

dispersed and well bounded. Later, the authors prepared a BC/graphene/PANI ternary system by LbL in situ culture method. The BC/G hydrogel was prepared as before and then deposited with polyaniline. The most promising material obtained presented 645 F g^{-1} of specific capacitance and retention of 82.2% after 1000 cycles [80].

A different type of configuration was proposed by Lv et al. [75] with a ternary system consisting of CNFs/molybdenum disulphide (MoS$_2$)/rGO hybrid aerogel. The aerogel was obtained by supercritical CO$_2$ drying and then compressed to a film to be used as an electrode. The MoS$_2$ provided high specific surface area to the electrode, while the GO enhanced the specific capacitance. The specific capacitance is about 916.42 F g^{-1}. Moreover, the capacity retention is more than 98% after 5000 charge/discharge cycles.

Selvam et al. [81] produced a wearable and flexible electronic device by single-step hydrothermal technique. For the purpose, GO, CH and copper chloride were cross-linked under an IL medium in an autoclave, forming a gel. The amine groups of CH can form strong interactions with the oxygen functionalities of GO and form dative bonds with copper metal ions. The temperature of the hydrothermal process (50, 75 and 100 °C) influenced the homogeneity of the hybrid material. The maximum specific capacitance achieved by the dried gel was 356 F g^{-1} using 100 °C in the hydrothermal step. This type of supercapacitor showed excellent long-term cyclic stability and retained its initial capacitance up to 200,000 cycles.

Table 4.4 gathers the information concerning polysaccharides/GBMs hybrids for energy applications.

References

1. Raimond JM, Brune M, Computation Q, De Martini F, Monroe C, Moehring DL, Knight PL, Plenio MB, Vedral V, Polzik ES, Variables C, Braunstein SL, Pati AK, Lukin MD, Cirac IJ, Zoller P, Han C, Xue P, Guo GC, Polyakov SV, Kuzmich A, Kimble HJ, Cirac JI, Kennedy TAB, Horodecki P, Horodecki R, Divincenzo DP, Smolin JA, Beige A, Kwek LC, Kok P, Sauer JA, You L, Zangwill A, Chapman MS, Nielsen M. Electric field effect in atomically thin carbon films. Science. 2004;20–23.
2. International union of pure and applied chemistry. Recommended Terminol. 1995;67:473–506
3. Bianco A, Cheng HM, Enoki T, Gogotsi Y, Hurt RH, Koratkar N, Kyotani T, Monthioux M, Park CR, Tascon JMD, Zhang J. All in the graphene family—a recommended nomenclature for two-dimensional carbon materials. Carbon. 2013;65:1–6.
4. Zhu Y, Murali S, Cai W, Li X, Suk JW, Potts JR, Ruoff RS. Graphene and graphene oxide: synthesis, properties, and applications. Adv Mater. 2010;22:3906–24.
5. Gonçalves G, Marques PAAP, Barros-Timmons A, Bdkin I, Singh MK, Emami N, Grácio J. Graphene oxide modified with PMMA via ATRP as a reinforcement filler. J Mater Chem. 2010;20:9927.
6. Marques PAAP, Gonçalves G, Cruz S, Almeida N, Singh MK, Grácio J, Sousa AACM. Functionalized graphene nanocomposites. In: Hashim AA, editors. Advances in nanocomposite technology, IntechOpen; 2011. p. 247–72.
7. Dreyer DR, Park S, Bielawsk CW, Ruoff RS. The chemistry of graphene oxide. Chem Soc Rev. 2010;39:228–40.
8. Acik M, Lee G, Mattevi C, Chhowalla M, Cho K, Chabal YJ. Unusual infrared-absorption mechanism in thermally reduced graphene oxide. Nat Mater. 2010;9:840–5.

9. Ege D, Kamali AR, Boccaccini AR. Graphene oxide/polymer-based biomaterials. Adv Eng Mater. 2017;19:16–34.
10. Feng H, Cheng R, Zhao X, Duan X, Li J. A low-temperature method to produce highly reduced graphene oxide. Nat Commun. 2013;4:1537–9.
11. Wick P, Louw-Gaume AE, Kucki M, Krug HF, Kostarelos K, Fadeel B, Dawson KA, Salvati A, Vázquez E, Ballerini L, Tretiach M, Benfenati F, Flahaut E, Gauthier L, Prato M, Bianco A. Classification framework for graphene-based materials. Angew Chemie Int Ed. 2014;53:7714–8.
12. Papageorgiou DG, Kinloch IA, Young RJ. Mechanical properties of graphene and graphene-based nanocomposites. Prog Mater Sci. 2017;90:75–127.
13. Li A, Zhang C, Zhang YF. Thermal conductivity of graphene-polymer composites: mechanisms, properties, and applications. Polymers. 2017;9:1–17.
14. Atif R, Shyha I, Inam F. Mechanical, thermal, and electrical properties of graphene-epoxy nanocomposites—a review. Polymers. 2016;8:281.
15. Terzopoulou Z, Kyzas GZ, Bikiaris DN. Recent advances in nanocomposite materials of graphene derivatives with polysaccharides. Materials. 2015;8:652–83.
16. Włodarczyk D, Urban M, Strankowski M. Chemical modifications of graphene and their influence on properties of polyurethane composites: a review. Phys Scr. 2016;91:104003.
17. Wang M, Duan X, Xu Y, Duan X. Functional three-dimensional graphene/polymer composites. ACS Nano. 2016;10:7231–47.
18. Wasupalli GK, Verma D. Polysaccharides as biomaterials. In: Thomas S, Balakrishnan P, Sreekala MS, editors. Fundamental biomaterials: polymers. Woodhead Publishing; 2018. p. 37–70.
19. Reina G, González-Domínguez JM, Criado A, Vázquez E, Bianco A, Prato M. Promises, facts and challenges for graphene in biomedical applications, Chem Soc Rev. 2017;4400–4416
20. Ramani D, Sastry TP. Bacterial cellulose-reinforced hydroxyapatite functionalized graphene oxide: a potential osteoinductive composite. Cellulose. 2014;21:3585–95.
21. Ciolacu DE, Suflet DM. Cellulose-based hydrogels for medical/pharmaceutical applications. In: Popa V, Volf I, editors. Biomass as renewable raw material to obtain bioproducts of high-tech value. Elsevier; 2018 p. 401–39.
22. Derivatives PP, Puiggal J. Hydrogels for biomedical applications: cellulose, chitosan, and protein/peptide derivatives. Gels. 2017;3:27.
23. Yu P, Bao R, Shi X, Yang W, Yang M. Self-assembled high-strength hydroxyapatite/graphene oxide/chitosan composite hydrogel for bone tissue engineering. Carbohydr Polym. 2017;155:507–15.
24. Stagnaro P, Schizzi I, Utzeri R, Marsano E, Castellano M. Alginate-polymethacrylate hybrid hydrogels for potential osteochondral tissue regeneration. Carbohydr Polym. 2018;185:56–62.
25. Mohan N, Mohanan PV, Sabareeswaran A, Nair P. Chitosan-hyaluronic acid hydrogel for cartilage repair. Int J Biol Macromol. 2017;104:1936–45.
26. Lee KJ, Il Yun S. Nanocomposite hydrogels based on agarose and diphenylalanine, Polymer. 2018;67:86–97.
27. Vedadghavami A, Minooei F, Mohammadi MH, Khetani S, Rezaei Kolahchi A, Mashayekhan S, Sanati-Nezhad A. Manufacturing of hydrogel biomaterials with controlled mechanical properties for tissue engineering applications. Acta Biomater. 2017;62:42–63.
28. Mohd N, Draman SFS, Salleh MSN, Yusof NB. Dissolution of cellulose in ionic liquid: a review. AIP Conf Proc. 2017;1809:020035.
29. Swatloski RP, Spear SK, Holbrey JD, Rogers RD. Dissolution of cellulose with ionic liquids. J Am Chem Soc. 2002;124:4974–5.
30. Xu M, Huang Q, Wang X, Sun R. Highly tough cellulose/graphene composite hydrogels prepared from ionic liquids. Ind Crop Prod. 2015;70:56–63.
31. Fan L, Yi J, Tong J, Zhou X, Ge H, Zou S, Wen H, Nie M. Preparation and characterization of oxidized konjac glucomannan/carboxymethyl chitosan/graphene oxide hydrogel. Int J Biol Macromol. 2016;91:358–67.

32. Shao W, Liu H, Liu X, Wang S, Zhang R. Anti-bacterial performances and biocompatibility of bacterial cellulose/graphene oxide composites. RSC Adv. 2015;5:4795–803.
33. Nandgaonkar AG, Wang Q, Fu K, Krause WE, Wei Q, Gorga R, Lucia LA. A one-pot biosynthesis of reduced graphene oxide (RGO)/bacterial cellulose (BC) nanocomposites. Green Chem. 2014;16:3195–201.
34. Mura S, Nicolas J, Couvreur P. Stimuli-responsive nanocarriers for drug delivery. Nat Mater. 2013;12:991–1003.
35. Deng L, Li Q, Al-Rehili S, Omar H, Almalik A, Alshamsan A, Zhang J, Khashab NM. Hybrid iron oxide-graphene oxide-polysaccharides microcapsule: a micro-matryoshka for on-demand drug release and antitumor therapy in vivo. ACS Appl Mater Interfaces. 2016;8:6859–68.
36. Luo H, Ao H, Li G, Li W, Xiong G, Zhu Y, Wan Y. Bacterial cellulose/graphene oxide nanocomposite as a novel drug delivery system. Curr Appl Phys. 2017;17:249–54.
37. Wang C, Wu C, Zhou X, Han T, Xin X, Wu J, Zhang J, Guo S. Enhancing cell nucleus accumulation and dna cleavage activity of anti-cancer drug via graphene quantum dots. Sci. Rep. 2013;3:1–8.
38. Schroeder KL, Goreham RV, Nann T. Graphene quantum dots for theranostics and bioimaging. Pharm Res. 2016;33:2337–57.
39. Javanbakht S, Namazi H. Doxorubicin loaded carboxymethyl cellulose/graphene quantum dot nanocomposite hydrogel films as a potential anticancer drug delivery system. Mater Sci Eng, C. 2018;87:50–9.
40. Justin R, Román S, Chen D, Tao K, Geng X, Grant RT, MacNeil S, Sun K, Chen B. Biodegradable and conductive chitosan–graphene quantum dot nanocomposite microneedles for delivery of both small and large molecular weight therapeutics. RSC Adv. 2015;5:51934–46.
41. Aggarwal P, Johnston CR. Geometrical effects in mechanical characterizing of microneedle for biomedical applications. Sensors Actuators B Chem. 2004;102:226–34.
42. Olszowska K, Pang J, Wrobel PS, Zhao L, Ta HQ, Liu Z, Trzebicka B, Bachmatiuk A, Rummeli MH. Three-dimensional nanostructured graphene: synthesis and energy, environmental and biomedical applications. Synth Met. 2017;234:53–85.
43. Henriques B, Gonçalves G, Emami N, Pereira E, Vila M, Marques PAAP. Optimized graphene oxide foam with enhanced performance and high selectivity for mercury removal from water. J Hazard Mater. 2016;301:453–61.
44. Mahfoudhi N, Boufi S. Nanocellulose as a novel nanostructured adsorbent for environmental remediation: a review. Cellulose. 2017;24:1171–97.
45. Crini G. Recent developments in polysaccharide-based materials used as adsorbents in wastewater treatment. Prog Polym Sci. 2005;30:38–70.
46. Brion-Roby R, Gagnon J, Deschênes JS, Chabot B. Development and treatment procedure of arsenic-contaminated water using a new and green chitosan sorbent: Kinetic, isotherm, thermodynamic and dynamic studies. Pure Appl Chem. 2018;90:63–77.
47. Bertoni FA, González JC, García SI, Sala LF, Bellú SE. Application of chitosan in removal of molybdate ions from contaminated water and groundwater. Carbohydr Polym. 2018;180:55–62.
48. Shen Y, Zhu X, Chen B. Size effects of graphene oxide nanosheets on the construction of three-dimensional graphene-based macrostructures as adsorbents. J Mater Chem A. 2016;4:12106–18.
49. Xiao J, Lv W, Song Y, Zheng Q. Graphene/nanofiber aerogels: performance regulation towards multiple applications in dye adsorption and oil/water separation. Chem Eng J. 2018;338:202–10.
50. Mi HY, Jing X, Politowicz AL, Chen E, Huang HX, Turng LS. Highly compressible ultra-light anisotropic cellulose/graphene aerogel fabricated by bidirectional freeze drying for selective oil absorption. Carbon. 2018;132:199–209.
51. Wei X, Huang T, Hui Yang J, Zhang N, Wang Y, Wan Zhou Z. Green synthesis of hybrid graphene oxide/microcrystalline cellulose aerogels and their use as superabsorbents. J Hazard Mater. (2017);335:28–38.

52. Yao Q, Fan B, Xiong Y, Jin C, Sun Q, Sheng C. 3D assembly based on 2D structure of cellulose nanofibril/graphene oxide hybrid aerogel for adsorptive removal of antibiotics in water. Sci Rep. 2017;7:1–13.
53. Sitko R, Musielak M, Zawisza B, Talik E, Gagor A. Graphene oxide/cellulose membranes in adsorption of divalent metal ions. RSC Adv. 2016;6:96595–605.
54. Xu L, Wang J. The application of graphene-based materials for the removal of heavy metals and radionuclides from water and wastewater. Crit Rev Environ Sci Technol. 2017;47:1042–105.
55. Chen X, Zhou S, Zhang L, You T, Xu F. Adsorption of heavy metals by graphene oxide/cellulose hydrogel prepared from NaOH/urea aqueous solution. Materials. 2016;9:582.
56. Khulbe KC, Matsuura T. Removal of heavy metals and pollutants by membrane adsorption techniques. Appl Water Sci. 2018;8:19.
57. Kyzas GZ, Travlou NA, Deliyanni EA. The role of chitosan as nanofiller of graphite oxide for the removal of toxic mercury ions. Colloids Surfaces B Biointerfaces. 2014;113:467–76.
58. Debnath S, Maity A, Pillay K. Magnetic chitosan–GO nanocomposite: Synthesis, characterization and batch adsorber design for Cr(VI) removal. J Environ Chem Eng. 2014;2:963–73.
59. Xu C, Shi L, Guo L, Wang X, Wang X, Lian H. Fabrication and characteristics of graphene oxide/nanocellulose fiber/poly(vinyl alcohol) film. J Appl Polym Sci. 2017;134:45345.
60. Yan N, Capezzuto F, Lavorgna M, Buonocore GG, Tescione F, Xia H, Ambrosio L. Borate cross-linked graphene oxide-chitosan as robust and high gas barrier films. Nanoscale. 2016;8:10783–91.
61. Ionita M, Pandele MA, Iovu H. Sodium alginate/graphene oxide composite films with enhanced thermal and mechanical properties. Carbohydr Polym. 2013;94:339–44.
62. Ashori A. Effects of graphene on the behavior of chitosan and starch nanocomposite films. Polym Eng Sci. 2014;54:2258–63.
63. Ahmed J, Mulla M, Arfat YA. Mechanical, thermal, structural and barrier properties of crab shell chitosan/graphene oxide composite films, Food Hydrocoll. (2017);71:141–148.
64. El Achaby M, Essamlali Y, El Miri N, Snik A, Abdelouahdi K, Fihri A, Zahouily M, Solhy A. Graphene oxide reinforced chitosan/polyvinylpyrrolidone polymer bio-nanocomposites. J Appl Polym Sci. 2014;131
65. Ma T, Chang PR, Zheng P, Ma X. The composites based on plasticized starch and graphene oxide/reduced graphene oxide. Carbohydr Polym. 2013;94:63–70.
66. Dutta S, Kim J, Ide Y, Ho Kim J, Hossain MSA, Bando Y, Yamauchi Y, Wu KC-W. 3D network of cellulose-based energy storage devices and related emerging applications. Mater. Horiz. 2017;4:522–45.
67. Wang Z, Tammela P, Strømme M, Nyholm L. Cellulose-based supercapacitors: material and performance considerations. Adv. Energy Mater. 2017;7:1700130.
68. Pérez-Madrigal MM, Edo MG, Aleman C. Powering the future: Application of cellulose-based materials for supercapacitors. Green Chem. 2016;18:5930–56.
69. Du X, Zhang Z, Liu W, Deng Y. Nanocellulose-based conductive materials and their emerging applications in energy devices—a review. Nano Energy. 2017;35:299–320.
70. Koga H, Tonomura H, Nogi M, Suganuma K, Nishina Y. Fast, scalable, and eco-friendly fabrication of an energy storage paper electrode. Green Chem. 2016;18:1117–24.
71. Zhang C, Cha R, Yang L, Mou K, Jiang X. Fabrication of cellulose/graphene paper as a stable-cycling anode materials without collector. Carbohydr Polym. 2018;184:30–6.
72. Kang Y-R, Li Y-L, Hou F, Wen Y-Y, Su D. Fabrication of electric papers of graphene nanosheet shelled cellulose fibres by dispersion and infiltration as flexible electrodes for energy storage. Nanoscale. 2012;4:3248–53.
73. Sevilla M, Ferrero GA, Fuertes AB. Graphene-cellulose tissue composites for high power supercapacitors. Energy Storage Mater. 2016;5:33–42.
74. Ma L, Liu R, Liu L, Wang F, Niu H, Huang Y. Facile synthesis of Ni(OH)2/graphene/bacterial cellulose paper for large areal mass, mechanically tough and flexible supercapacitor electrodes. J Power Sources. 2016;335:76–83.
75. Lv Y, Li L, Zhou Y, Yu M, Wang J, Liu J, Zhou J. A cellulose-based hybrid 2D material aerogel for a flexible all-solid-state supercapacitor with high specific capacitance. RSC Adv. 2017;7:43512–20.

76. Liu Y, Zhou J, Zhu E, Tang J, Liu X, Tang W. Facile synthesis of bacterial cellulose fi bres covalently intercalated with graphene oxide by. J Mater Chem C Mater Opt Electron Devices. 2015;3:1011–1017

77. Liu Y, Zhou J, Tang J, Tang W. Three-dimensional, chemically bonded polypyrrole/bacterial cellulose/graphene composites for high-performance supercapacitors. Chem Mater. 2015;27:7034–41.

78. Gao K, Shao Z, Li J, Wang X, Peng X, Wang W, Wang F. Cellulose nanofiber–graphene all solid-state flexible supercapacitors. J Mater Chem A. 2013;1:63–7.

79. Liu R, Ma L, Huang S, Mei J, Xu J, Yuan G. Large areal mass, flexible and freestanding polyaniline/bacterial cellulose/graphene film for high-performance supercapacitors. RSC Adv. 2016;6:107426–32.

80. Luo H, Dong J, Zhang Y, Li G, Guo R, Zuo G, Ye M, Wang Z, Yang Z, Wan Y. Constructing 3D bacterial cellulose/graphene/polyaniline nanocomposites by novel layer-by-layer in situ culture toward mechanically robust and highly flexible freestanding electrodes for supercapacitors. Chem Eng J. 2018;334:1148–58.

81. Selvam S, Balamuralitharan B, Jegatheeswaran S, Kim M-Y, Karthick SN, Anandha Raj J, Boomi P, Sundrarajan M, Prabakar K, Kim H-J. Electrolyte-imprinted graphene oxide-chitosan chelate with copper crosslinked composite electrodes for intense cyclic-stable, flexible supercapacitors. J Mater Chem A. 2017;5:1380–6.

82. Andreeva TD, Stoichev S, Taneva SG, Krastev R. Hybrid graphene oxide/polysaccharide nanocomposites with controllable surface properties and biocompatibility. Carbohydr Polym. 2018;181:78–85.

83. Saravanan S, Chawla A, Vairamani M, Sastry TP, Subramanian KS, Selvamurugan N. Scaffolds containing chitosan, gelatin and graphene oxide for bone tissue regeneration in vitro and in vivo. Int J Biol Macromol. 2017;104:1975–85.

84. Serrano-Aroca Á, Iskandar L, Deb S. Green synthetic routes to alginate-graphene oxide composite hydrogels with enhanced physical properties for bioengineering applications. Eur Polym J. 2018;103:198–206.

85. Rajesh R, Ravichandran YD. Development of new graphene oxide incorporated tricomponent scaffolds with polysaccharides and hydroxyapatite and study of their osteoconductivity on MG-63 cell line for bone tissue engineering. RSC Adv. 2015;5:41135–43.

86. Pal N, Dubey P, Gopinath P, Pal K. Combined effect of cellulose nanocrystal and reduced graphene oxide into poly-lactic acid matrix nanocomposite as a scaffold and its anti-bacterial activity. Int J Biol Macromol. 2017;95:94–105.

87. Sun H, Zhang L, Xia W, Chen L, Xu Z, Zhang W. Fabrication of graphene oxide-modified chitosan for controlled release of dexamethasone phosphate. Appl Phys A Mater Sci Process. 2016;122:1–8.

88. Liu X, Cheng X, Wang F, Feng L, Wang Y, Zheng Y, Guo R. Targeted delivery of SNX-2112 by polysaccharide-modified graphene oxide nanocomposites for treatment of lung cancer. Carbohydr Polym. 2018;185:85–95.

89. Li Y, Zhang H, Fan M, Zheng P, Zhuang J, Chen L. A robust salt-tolerant superoleophobic alginate/graphene oxide aerogel for efficient oil/water separation in marine environments. Sci. Rep. 2017;7:1–7.

90. Suo F, Xie G, Zhang J, Li J, Li C, Liu X, Zhang Y, Ma Y, Ji M. A carbonised sieve-like corn straw cellulose–graphene oxide composite for organophosphorus pesticide removal. RSC Adv. 2018;8:7735–43.

91. Yang A, Wu J, Huang CP. Graphene oxide-cellulose composite for the adsorption of uranium (VI) from dilute aqueous solutions. 2018;22:1–7

92. Yakout AA, El-Sokkary RH, Shreadah MA, Abdel OG. Hamid, Cross-linked graphene oxide sheets via modified extracted cellulose with high metal adsorption. Carbohydr Polym. 2017;172:20–7.

93. Algothmi WM, Bandaru NM, Yu Y, Shapter JG, Ellis AV. Alginate–graphene oxide hybrid gel beads: An efficient copper adsorbent material. J Colloid Interface Sci. 2013;397:32–8.

94. Huang H-D, Liu C-Y, Li D, Chen Y-H, Zhong G-J, Li Z-M. Ultra-low gas permeability and efficient reinforcement of cellulose nanocomposite films by well-aligned graphene oxide nanosheets. J. Mater. Chem. A. 2014;2:15853–63.

95. El Achaby M, El Miri E, Snik A, Zahouily M, Abdelouahdi K, Fihri A, Barakat A, Solhy A. Mechanically strong nanocomposite films based on highly filled carboxymethyl cellulose with graphene oxide J Appl Polym Sci. 2015;133

96. Huang Q, Xu M, Sun R, Wang X. Large scale preparation of graphene oxide/cellulose paper with improved mechanical performance and gas barrier properties by conventional papermaking method. Ind Crops Prod. 2016;85:198–203.

97. Lee DB, Kim DW, Shchipunov Y, Ha C-S. Effects of graphene oxide on the formation, structure and properties of bionanocomposite films made from wheat gluten with chitosan. Polym Int. 2016;65:1039–45.

98. Demitri C, De Benedictis VM, Madaghiele M, Corcione CE, Maffezzoli A. Nanostructured active chitosan-based films for food packaging applications: effect of graphene stacks on mechanical properties. Meas. J. Int. Meas. Confed. 2016;90:418–23.

99. Ge X, Li H, Wu L, Li P, Mu X, Jiang Y. Improved mechanical and barrier properties of starch film with reduced graphene oxide modified by SDBS. J Appl Polym Sci. 2017;134:44910.

100. Jose J, Al-Harthi MA. Citric acid crosslinking of poly(vinyl alcohol)/starch/graphene nanocomposites for superior properties. Iran Polym J. 2017;26:579–87.

101. Aqlil M, Moussemba Nzenguet A, Essamlali Y, Snik A, Larzek M, Zahouily M. Graphene oxide filled lignin/starch polymer bionanocomposite: structural, physical, and mechanical studies. J Agric Food Chem. 2017;65:10571–81.

102. Bher A, Unalan IU. Toughening of poly(lactic acid) and thermoplastic cassava starch reactive blends using graphene nanoplatelets. Polymers. 2018;10:1–18.

103. Nie L, Liu C, Wang J, Shuai Y, Cui X, Liu L. Effects of surface functionalized graphene oxide on the behavior of sodium alginate. Carbohydr Polym. 2015;117:616–23.

104. Zheng Q, Cai Z, Ma Z, Gong S. Cellulose nanofibril/reduced graphene oxide/carbon nanotube hybrid aerogels for highly flexible and all-solid-state supercapacitors. ACS Appl Mater Interfaces. 2015;7:3263–71.

105. Wang Y, He Z, Wang Y, Fan C, Liu C, Peng Q, Chen J, Feng Z. Preparation and characterization of flexible lithium iron phosphate/graphene/cellulose electrode for lithium ion batteries. J Colloid Interface Sci. 2018;512:398–403.

106. Chang Y, Zhou L, Xiao Z, Liang J, Kong D, Li Z, Zhang X, Li X, Zhi L. Embedding reduced graphene oxide in bacterial cellulose-derived carbon nanofibril networks for supercapacitors. ChemElectroChem. 2017;4:2448–52.

107. Zhang Y, Wang F, Zhang D, Chen J, Zhu H, Zhou L, Chen Z. New type multifunction porous aerogels for supercapacitors and absorbents based on cellulose nanofibers and graphene. Mater Lett. 2017;208:73–6.

108. Sun G, Li B, Ran J, Shen X, Tong H. Three-dimensional hierarchical porous carbon/graphene composites derived from graphene oxide-chitosan hydrogels for high performance supercapacitors. Electrochim Acta. 2015;171:13–22.

109. Zhang Y, Zhu J-Y, Ren H-B, Bi Y-T, Zhang L. Facile synthesis of nitrogen-doped graphene aerogels functionalized with chitosan for supercapacitors with excellent electrochemical performance. Chinese Chem. Lett. 2017;28:935–42.

110. Hosseini MG, Shahryari E. A Novel High-performance supercapacitor based on chitosan/graphene oxide-mwcnt/polyaniline. J Colloid Interface Sci. 2017;496:371–81.

111. Wu W, Li Y, Yang L, Ma Y, Pan D, Li Y. A facile one-pot preparation of dialdehyde starch reduced graphene oxide/polyaniline composite for supercapacitors. Electrochim Acta. 2014;139:117–26.

112. Ma T, Chang PR, Zheng P, Zhao F, Ma X. Porous graphene gels: preparation and its electrochemical properties. Mater Chem Phys. 2014;146:446–51.

Chapter 5
Polysaccharides-Based Hybrids with Carbon Nanotubes

Carbon nanotubes (CNTs) are nanometric scale sp^2 carbon bonded materials with a tube-like structure formed by rolling-up graphene sheet(s) in a seamless way into a cylinder with open or closed ends [1–3]. This graphene allotrope can be classified into single-walled (SWCNT) and multi-walled (MWCNT) carbon nanotubes depending on the number of rolled-up graphene layers. The former was discovered in 1993 [4] and consists of single graphene sheet rolled into a cylinder, whereas the latter was discovered in 1991 [5] and is formed by multiple stacked graphene layers in the form of cylinders with an interlayer spacing of 0.34 nm [6–8], as illustrated in Fig. 5.1. The diameters of SWCNTs and MWCNTs are typically in the range of 0.2–2.0 and 2–100 nm, respectively [9], while the length varies from less than 100 nm to several centimetres [1]. These differences translate into materials with different physical and chemical properties. The most remarkable features of CNTs comprise their high surface area, aspect ratio, thermal conductivity, electron mobility and mechanical strength [2]. In fact, CNTs are stronger than steel with a tensile strength of 150–180 GPa and tensile modulus between 640 GPa and 1 TPa, as well as better conductors than copper with electrical conductivity values ranging from 10^7–10^8 S m^{-1} [3, 6, 8]. In terms of thermal conductivity, CNTs surpass diamond with SWCNTs attaining a value of 3500 W m^{-1} K^{-1} at room temperature [1, 10]. Nevertheless, the main downside of this carbonaceous nanomaterial is associated with its hydrophobicity, and, consequently, poor processability and compatibility with other materials, which can be circumvented by functionalization with different functional groups (*e.g.*, hydroxyl, carboxyl and amine moieties) [6].

CNTs are mainly produced using graphite as carbon source via various methods such as arc-discharge, laser ablation and chemical vapour deposition [7, 11]. Additionally, CNTs can be fabricated from renewable feedstocks, including vegetable oils, plant derivatives, and other types of biomasses [12], by using conventional synthesis methodologies [11]. Worth noting is the fact that the CNTs market size was $3.43 billion in 2016 and is projected to worth $8.7 billion by 2022, according to the report published by Research and Markets [13]. These carbon-based nanomaterials have been widely studied in almost all domains of modern science and technology,

C. Vilela et al., *Polysaccharide Based Hybrid Materials*, Biobased Polymers, https://doi.org/10.1007/978-3-030-00347-0_5

(a)   **(b)**

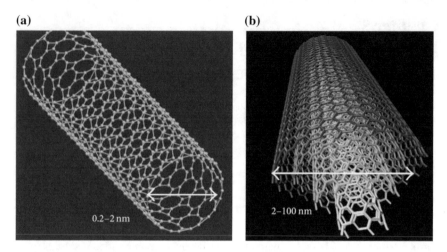

Fig. 5.1 a Single-walled carbon nanotubes (SWCNTs) and **b** multi-walled carbon nanotubes (MWCNTs). Reprinted with permission from Ref. [9]

including in biosensing [6], heavy metal adsorption [14], energy conversion and storage [2], filtration membranes [15], cancer targeting and drug delivery [9, 16], just to mention a few applications. Furthermore, CNTs are also attractive materials as nanofillers for composites [17], particularly for metal-matrix [8], polymer-matrix [7, 18] and cement-matrix [19] nanocomposites, due to their intrinsically high mechanical performance. Despite the abundant literature on the topic as evidenced by the countless review papers enumerated in the previous paragraphs, none is devoted solely to polysaccharides/CNTs hybrid materials. Thus, the emphasis of the publications surveyed in this chapter will be on hybrid materials based on CNTs and polysaccharides, namely cellulose, chitin, chitosan (CH) and starch, that privileged novelty and potential applications.

5.1 Cellulose/CNTs Hybrid Materials

Cellulose has been broadly investigated in combination with CNTs, as corroborated by the extensive list of publications on the topic with a considerable number of studies dealing with cellulose [20–22] and its derivatives, such as cellulose acetate [23, 24], carboxymethyl cellulose (CMC) [25], and regenerated cellulose [26, 27], as well as with the nanoscale forms of cellulose, namely cellulose nanofibrils (CNFs) [28, 29], cellulose nanocrystals (CNCs) [30, 31] and bacterial cellulose (BC) [32, 33], to fabricate hybrid materials in the form of films [30, 34], membranes [31, 35], aerogels [20, 36], hydrogels [25] and fibres [37], as summarized in Table 5.1. These cellulose/CNTs-based hybrid materials have showed tremendous potential for application in multiple domains, for example, as electrically conductive aerogels [20], aerogels for vapour sensing [36], water sensors [21], multifunctional sensing applications [34, 38], amperometric sensing probes [24], electrodes for supercapacitors [39], water filtration [35], mixed matrix membranes for CO_2/N_2 separation [23], transdermal device for drug delivery [25], scaffold for bone regeneration [33], and gauzes for haemostatic applications [27].

Within the combined contexts of novelty and application, electrically conductive aerogels based on cellulose cotton linters and MWCNTs (3, 5 and 10 wt%) were prepared for application in vapour sensing of volatile organic compounds (VOCs) [36]. These aerogels with highly porous networks exhibit rapid response, high sensitivity and good reproducibility to both polar and nonpolar VOCs such as methanol, ethanol, acetone, chloroform, tetrahydrofuran, toluene and hexane; hence, can be used as reproducible sensors for VOCs as well as for other gases analysis at room temperature [36].

In a different study, Zeng et al. [28] described the fabrication of flexible dielectric papers based on CNFs and CNTs (0.5–4.5 wt%) via a facile vacuum-assisted self-assembly technique for dielectric energy storage. These homogeneous, highly ordered and degradable papers are mechanically flexible and present good mechanical strength (Young's modulus > 5 GPa) and dielectric energy storage capability $(0.81 \pm 0.1 \, \text{J cm}^{-3})$ [28]. Despite the fairly good results attained in this study, papers with better mechanical performances were expectable given the inclusion of CNTs as a carbonaceous filler.

In the environmental domain, Ahmad et al. [23] reported the development of mixed matrix membranes from cellulose acetate (CA) and MWCNTs functionalized with β-cyclodextrins for carbon dioxide (CO_2)/nitrogen (N_2) separation. The enhanced permeance and selectivity of these membranes towards the separation of both CO_2 and N_2 with a threshold amount of 0.1 wt% of functionalized MWCNTs make them a promising means for CO_2 capture, a strategy that in being tackled to reduce the Earth's greenhouse effect [23].

In the health domain, an important publication includes the work of Gutiérrez-Hernández et al. [33] about the design of scaffolds for bone regeneration based on BC and MWCNTs–COOH. The inclusion of MWNTs–COOH (2.5 and 5.0 wt%) onto the BC matrix enhanced the mechanical performance of the scaffolds (the storage modulus was threefold as compared to pristine BC), and favoured osteoblastic cell spreading, adhesion, and proliferation owing to the interfacial compatibility of the cells with the scaffolds. Therefore, these three-dimensional hybrid scaffolds are suitable for osteoblastic cell culture and, concomitantly, for bone regeneration [33].

Recently, Cheng et al. [27] developed MWCNTs/cellulose gauzes for haemostatic applications. According to this study, unmodified (*i.e.* MWCNTs) and functionalized MWCNTs (i.e. MWCNTs–NH$_2$ and MWCNTs–COOH) were grafted to oxidized-regenerated-cellulose gauze, originating hybrid materials with augmented haemostatic performance. In fact, the oxidized regenerated cellulose gauze containing MWCNTs–COOH showed the lowest haemostatic efficiency with about 207 and 296 s, which translates into a significant reduction of the bleeding time on rabbit ear artery and liver haemorrhage model, respectively [27].

Table 5.1 Examples of polysaccharides-based hybrids with CNTs, the preparation methodologies and potential applications

Polysaccharide	Type of CNTs	Methodology	Application	References
Cellulose	MWCNTs	CNTs were mixed with cotton linters, followed by freeze-drying	Electrically conductive aerogels	[20]
	MWCNTs	CNTs were mixed with cotton linters prior to casting and coagulation	Electrically conductive films for use as water sensors	[21]
	MWCNTs	CNTs were mixed with cotton linters, followed by freeze-drying	Electrically conductive aerogels for vapour sensing	[36]
	MWCNTs	CNTs were mixed with cotton linters prior to wet spinning process	Electrically conductive fibres with potential to be used as wearable electronics	[37]
	MWCNTs	CNTs were mixed with cellulose pulp, followed by suction filtration method for paper fabrication	Electrically conductive papers with potential for batteries and supercapacitors	[22]
	MWCNTs MWCNTs–OH	CNTs were mixed with cellulose microfibres pulp, before paper handsheets fabrication	Smart papers for multifunctional sensing	[38]
CA	MWCNTs	CNTs functionalized with β-cyclodextrins were mixed with CA, prior to membrane formation by wet phase inversion	Mixed matrix membrane for CO_2/N_2 separation	[23]
	MWCNTs–COOH	CNTs were mixed with CA followed by phase inversion	Macroporous membranes for water filtration	[35]
	MWCNTs	CNTs were dispersed into CA polymer matrix, followed by casting onto a glassy carbon electrode	Amperometric catecholamines sensing probe	[24]

(continued)

Table 5.1 (continued)

Polysaccharide	Type of CNTs	Methodology	Application	References
CMC	MWCNTs–COOH	CNTs were mixed with CMC via ultrasonication followed by vacuum drying	Hybrid hydrogel for transdermal drug delivery	[25]
Regenerated cellulose	SWCNTs	CNTs were mixed with cellulose in the presence of an ionic liquid prior to casting and regeneration in water	Electrically conductive films	[26]
	MWCNTs MWCNTs–NH$_2$ MWCNTs–COOH	CNTs-coated oxidized gauzes were prepared by freeze-drying	Gauzes for hemostatic applications	[27]
CNFs	CNTs	CNTS were mixed with CNFs via sonication, followed by vacuum-assisted self-assembly technique	Flexible papers for dielectric energy storage	[28]
CNCs	MWCNTs	CNTs were mixed with CNCs followed by vacuum filtration and vacuum drying	Electrically conductive hybrid films with potential for sensing applications	[34]
	MWCNTs–OH	CNTs were mixed with a CNCs slurry prior to membrane fabrication by phase inversion	Nanocomposites with potential as water filtration membranes	[31]
BC	MWCNTs–COOH	CNTs were mixed with BC pulp and alginate/D-mannitol solution, followed by cross-linking and freeze-drying	Scaffolds for bone regeneration	[33]
	MWCNTs	CNTs were mixed with BC containing Pt precursors, followed by reduction with H$_2$ and membrane drying	Composite as anode catalyst for proton exchange membrane fuel cells	[32]

(continued)

Table 5.1 (continued)

Polysaccharide	Type of CNTs	Methodology	Application	References
Chitin	MWCNTs	CNTs were mixed with chitin flakes and magnetite nanoparticles using agate mortar	Magnetic hybrid for the removal of organic dyes (Rose Bengal) from solutions	[44]
	MWCNTs	CNTs suspension was mixed with chitin under ultrasonication, followed by filtration and drying	Gel-film for foldable conductive paper	[41]
	MWCNTs	CNTs grafted with poly(4-vinylpyridine) were mixed with chitin nano-whiskers by sonication-assisted assembly, followed by freeze-drying	Mesoporous aerogel with potential application as thermal insulators, catalyst supports, and biomedical materials	[40]
	MWNTs (coated with CMC)	CNTs were mixed with chitin, followed by coagulation and casting methods, and O_2 plasma treatment	Scaffolds for tissue engineering of neurons and, potentially, as an implantable electrode for stimulation and repair of neurons	[45]
	MWCNTs–COOH	CNTs were blended with chitin, prior to regeneration and freeze-drying	Hydrogels as neuronal growth substrates	[42]
	MWCNTs–COOH	CNTs were blended with chitin, followed by freezing/thawing method and lysine immobilization	Blood compatible bilirubin microspheres adsorbents	[43]
CH	MWCNTs–COOH	CNTs were functionalized with a CH-folic acid conjugate by liquid-gel transition	Nanoparticle hybrids as gene delivery materials	[50]
	MWCNTs–COOH	CNTs were mixed with CH followed by bead formation	Sorbent for removal of heavy metal ions from aqueous solutions	[52]

(continued)

Table 5.1 (continued)

Polysaccharide	Type of CNTs	Methodology	Application	References
	MWCNTs–COOH	CNTs were mixed with CH prior to electrophoretic deposition on a titanium substrate	Cell stimulation and therapeutics delivery for bone regenerating implants	[51]
	MWCNTs–COOH	CNTs were mixed with CH and silica, followed by sol-gel process, freeze-drying and hot pressing	Membranes with potential for guided bone regeneration	[57]
	MWCNTs	CNTs were blended with CH (or CNCs), followed by layer-by-layer assembly of thin films	Transparent conductive thin films with potential to fabricate biocompatible transparent electrodes	[58]
	CNTs–COOH	CNTs were mixed with CH before dripping onto liquid nitrogen, followed by freeze-drying	Spherical beads for bilirubin adsorption	[49]
	MWCNTs–COOH	CNTs were blended with cationic CH derivative prior to cross-linking, followed by casting method	Superhydrophobic and antibacterial hybrid membrane for application in food, bioengineering and medical fields	[59]
	MWCNTs	CNTs were mixed with CH before electrochemical deposition onto a microelectrode array	Microelectrode array for selective recognition of 5-hydroxytryptamine and dopamine molecules	[46]
	MWCNTs–COOH	CNTs were coated with silica through a sol-gel process, followed by blending with CH and casting	Polymer electrolyte membranes for fuel cells	[53]
	MWCNTs–COOH	CNTs mixed with CH were deposited onto titanium disks by electrophoretic deposition, followed by atom layer deposition of ZnO	Antibacterial hybrids nanostructures used as titanium implants	[60]

(continued)

Table 5.1 (continued)

Polysaccharide	Type of CNTs	Methodology	Application	References
	SWCNTs–COOH	CNTs were covalently bonded to CH followed by casting on the surface of a glass carbon electrode	Electrochemical biosensor for serum leptin detection	[47]
	MWCNTs	CNTs were mixed with CH followed by electrodeposition onto a polished Au electrode	Electrochemical biosensor for the in vivo monitoring of dopamine	[48]
	MWCNTs	CNTs were mixed with CH for ink preparation, followed by layer-by-layer deposition onto paper-based electrodes	Flexible paper-based electrodes for point of care and point of need testing	[56]
	MWCNTs	CNTs were coated with CH by surface-deposition, followed by blending and casting with CH matrix	Polymer electrolyte membranes for fuel cells	[54]
	MWCNTs–COOH	CNTs grafted with organic molecules (CNTs fluids) were mixed with CH matrix, followed by solution casting and cross-linking with sulfuric acid	Polymer electrolyte membranes for fuel cells	[55]
Starch	MWCNTs	CNTs were mixed with plasticized starch followed by casting method	Electrical conductive films	[67]
	MWCNTs	CNTs were mixed with plasticized starch prior to casting method	Gas barrier and electrical conductive films	[68]
	MWCNT–FeAl$_2$O$_4$	CNTs were mixed with plasticized starch before thermal gelatinization	Films with potential in the packaging industry	[61]
	MWCNTs–OH	CNTs were mixed with plasticized starch, followed by film casting	Films with potential for secondary packaging in the food sectors	[63]

(continued)

Table 5.1 (continued)

Polysaccharide	Type of CNTs	Methodology	Application	References
	CNTs–OH	CNTs were mixed with plasticized hydroxypropyl starch before thermal gelatinization	Films with potential for packaging and coating application	[62]
	MWCNTs–COOH	CNTs functionalized with vitamin C were mixed with plasticized starch via ultrasonication and casting method	Films for removal of organic dyes pollutants	[64]
	MWCNTs–COOH	CNTs functionalized with fructose were mixed with plasticized starch prior to casting method	Films with potential for the removal of dyes pollutants from wastewater	[65]
	MWCNTs–COOH	CNTs functionalized with glucose were mixed with plasticized starch followed by casting method	Nanocarrier for the delivery of hydrophobic drugs	[66]
ALG	SWCNTs–COOH	Freeze drying-mechanically pressing technique	Porous hybrid paper as flexible conductors and phase change materials	[69]
	MWCNTs–COOH	CNTs were incorporated in ALG by homogenization prior to gelation	Hydrogel substrates for cell culture applications, cell therapy and tissue engineering	[70]
	MWCNTs	CNTs homogenization with ALG via surfactant assisted dispersion before inducing the gelation with $BaCl_2$	Hybrid electrode for probing microbial electroactivity	[71]
ALG and CH	SWCNTs	Layer-by-layer assembly of ALG and CH on CNTs, followed by seed growth of gold nanoparticles	Photothermal therapy	[73]

(continued)

Table 5.1 (continued)

Polysaccharide	Type of CNTs	Methodology	Application	References
ALG and cellulose	MWCNTs	CNTs were blended with ALG, cellulose and ibuprofen, followed by ionic cross-linking	pH and electric field dual-stimulus responsive hybrid hydrogel for drug delivery	[72]
Carrageenan	MWCNTs–COOH	CNTs were blended with carrageenan and model molecule, followed by carrageenan gelation with KCl	Carriers for remotely activated drug delivery	[74]
	MWCNTs–COOH	CNTs were functionalized with carrageenan, followed by in situ synthesis of Fe_3O_4 nanoparticles	Magnetic adsorbent to remove cationic dyes from wastewaters	[75]
Hyaluronan	MWCNTs–COOH	CNTs were mixed with hyaluronan prior to the addition of cross-linker, followed by freeze-drying	Porous scaffolds to induce neural regeneration in tissues of the central or peripheral nervous system	[76]
	MWCNTs–NH$_2$	CNTs were mixed with hyaluronan and applied on the electrode surface, followed by antigen immobilization	Electrochemical immunosensor for hepatitis B	[77]

5.2 Chitin/CNTs Hybrid Materials

The combination between the marine polysaccharide chitin and CNTs has also received some attention to produce hybrid materials (Table 5.1) in the form of aerogels [40], hydrogels [41, 42], and microspheres [43], for diverse applications including absorbents for removal of dyes pollutants [44], conductive paper [41], blood purified therapy [43] and neuronal growth substrates [42, 45].

A representative example of the ongoing research in this topic deals with the use of chitin/CNTs-based scaffolds for neuron repair/regeneration [42, 45] since CNTs can promote cell adhesion, proliferation and differentiation of neuronal cells. Singh et al. demonstrated that the inclusion of MWCNTs (coated with carboxymethyl cellulose) into chitin, dissolved in ionic liquids, improved the electrical conductivity of the O_2 plasma-treated chitin/CNTs composite films [45]. In addition, these biocompatible and electrically-conductive hybrid materials showed increased neuron attachment and supported neural synapses, while maintaining their functional integrity, which indicates that the neurons remained functionally-active on the scaffolds even after 21 days of testing [45].

Wu et al. [42] created hybrid hydrogels based on chitin and oxidized MWCNTs also with potential as neuronal growth substrates for the peripheral nerve regeneration. These hydrogels with a compact and neat nanofibrillar network morphology displayed hemocompatibility and biocompatibility, as well as enhanced cellular proliferation and adhesion of nerve cells such as PC12 and RSC96 cells. Besides, these biocompatible chitin/CNTs hybrids are quite versatile materials in terms of forms and shapes since they can be fabricated as films, aerogels, fibres, macroporous hydrogels and nanofibrous microspheres for extended application in nerve tissue engineering [42].

In a different context, the same research group developed blood compatible bilirubin adsorbents based on chitin and CNTs [43]. The authors describe the fabrication of microspheres via dispersion of carboxylated MWCNTs in NaOH/urea aqueous solution in the presence of chitin, followed by thermal induction that yielded nanofibrous hybrid microspheres. The immobilization of lysine through cross-linking on the interconnected porous structure of the chitin/CNTs hybrids contributed to a higher efficiency of bilirubin adsorption from plasma because of the specific interactions between the amino moieties of lysine residues and the carboxyl groups of bilirubin. This combination of properties underlines the potential of these hybrid microspheres as a bilirubin adsorbent in hemoperfusion applications for treatment of liver diseases [43].

5.3 Chitosan/CNTs Hybrid Materials

Contrary to chitin, CH is being extensively studied for the development of polysaccharide-based hybrid materials (Table 5.1). In fact, one of the strongest potential of hybrid materials based on CH and CNTs lies in their myriad applications as, for example, microelectrode array for selective recognition of organic molecules [46], biosensors for serum leptin detection [47], in vivo monitoring of dopamine [48], bilirubin adsorption [49], gene delivery materials [50], cell stimulation and therapeutics delivery [51], sorbent for removal of heavy metal ions from aqueous solutions [52], polymer electrolyte membranes for fuel cells [53–55], flexible electrodes [56], among other applications [57–60].

A recent publication includes the work of Shukla et al. [48] concerning CH films loaded with MWCNTs that were electrodeposited onto a polished Au electrode. The authors reported that the inclusion of MWCNTs onto CH films affected the sensing performance of dopamine in the presence of biological interference (e.g. uric acid) by increasing the diffusion and electron transfer rate coefficients of the sensor. Moreover, the sensor with higher MWCNTs content offers better sensitivity ($3.00 \, \mu A \, L \, \mu mol^{-1}$ for 1.75% MWCNTs loading, versus $0.01 \, \mu A \, L \, \mu mol^{-1}$ for 1% loading) but an inferior limit-of-detection ($2.00 \, \mu mol \, L^{-1}$ versus $1.00 \, \mu mol \, L^{-1}$,

(a)

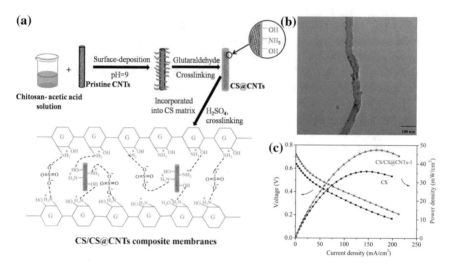

Fig. 5.2 a Synthesis process of chitosan-coated CNTs and the corresponding composite membranes, **b** TEM micrograph of chitosan-coated CNTs, and **c** polarization and power density curves of direct methanol fuel cell single cells tested of pure chitosan and composite membranes at 5 M methanol concentration at 70 °C. Reprinted with permission from Ref. [54]. Copyright 2017 John Wiley & Sons

respectively). Consequently, these films can be used as electrochemical biosensors for the in vivo monitoring of dopamine and of other redox-active molecules [48].

In a different approach, Figueredo et al. [56] describes the layer-by-layer construction of CH/MWCNTs flexible paper-based electrodes. These low-cost and disposable electrodes with good electron transfer and mechanical deformation endurance demonstrated great potential in the detection of Pb at trace levels in water samples in the presence of Bi (10–200 ppb) with a limit detection of 6.74 ppb, as well as dopamine in presence of uric and ascorbic acids with the limit of detection of 6.32 μM. Thus, these electrodes can be a versatile electrochemical platform for point-of-care and point-of-need testing, with potential for utilization in developing countries with limited resources and disperse population that are quite far from clinical and other analytical facilities [56].

Another relevant addition to the field of CH/CNTs hybrids includes the work of Ou et al. [54] about polymer electrolyte membranes prepared by blending and casting of CH matrix and MWCNTs functionalized with CH via a facile noncovalent surface-deposition and cross-linking method (Fig. 5.2a). These membranes present a proton conductivity of 34.6 mS cm^{-1} at 80 °C, which is about 1.5-fold of the conductivity of pure CH membrane. Furthermore, the direct methanol fuel cell performance of the membranes was evaluated by single cell, tested at 70 °C, and the membrane exhibits a peak power density of 47.5 mW cm^{-2} (Fig. 5.2c), which is higher than that of pristine CH [54].

As a last and noteworthy publication, Wang et al. [55] also developed proton exchange hybrid membranes for fuel cells applications via incorporation of solvent-free MWCNTs fluids with liquid-like behaviour at room temperature (prepared through an ion exchange method) onto the CH matrix by a solution casting method. It was observed that these organic-grafted MWCNTs with liquid-like behaviour improved simultaneously the interface compatibility and the mechanical performance of the membranes. Moreover, the proton transfer pathway provided by the interactions between the $-SO_3^-$ and $-NH_3^+$ groups of MWCNT fluids and CH, respectively, contributed to the maximum proton conductivity of 44 mS cm^{-1} at 80 °C and power density of 48.46 mW cm^{-2} [55], which are higher than the values reported by Ou et al. [54] in the example described above.

5.4 Starch/CNTs Hybrid Materials

Starch is another polysaccharide that has been combined with CNTs to generate organic-inorganic hybrid materials with improved functional properties for diverse applications (Table 5.1), including food packaging [61–63], removal of pollutant dyes from wastewater [64, 65] and drug delivery [66]. As an illustrative example, Cheng et al. [67] fabricated nanocomposite films based on plasticized-starch and modified-MWCNTs by casting method. The modified-MWCNTs were first oxidized by Hummer's method and then reduced by glucose, which originated CNTs with 15 and 8 wt% oxygen-containing groups, respectively. The incorporation of modified-MWCNTs nanofillers on the plasticized-starch matrix promoted a reinforcing effect and render the materials with conductive properties [67]. Along the same idea, Swain et al. [68] prepared nanocomposite films by solution casting method from plasticized starch and functionalized-MWCNTs. These films exhibited good conductive and gas barrier properties that increased with the increasing content of MWCNTs from 0.5 to 3.0 wt% [68].

In a different study, hercynite ($FeAl_2O_4$) nanoparticles anchored to the surface of MWCNTs were used as a nanohybrid filler to reinforce plasticized starch films via gelatinization with potential application in the packaging industry [61]. The inclusion of small amounts of functionalized-MWCNTs (0.04 wt%) triggered an augment of 370% in the Young's modulus, 138% in tensile strength and 350% in tensile toughness, as well as a 70% reduction in water vapor permeability relative to the plasticized starch matrix. These increments are probably a direct result of the modified MWCNTs homogeneous dispersion and affinity with the plasticizers [61].

More recently, Liu et al. [62] investigated the physicochemical properties changes from multi-scale structures of films based on modified-starch (hydroxypropyl derivative) and CNTs–OH (0.05–2.0%). According to this study, factors such as molecular interaction, short range molecular conformation, crystalline structure and aggregated structure affect the properties of the ensuing films, and therefore should be considered when designing starch-based nanocomposite films for packaging and coating applications. In fact, the increase of CNTs content led to the breakage of origi-

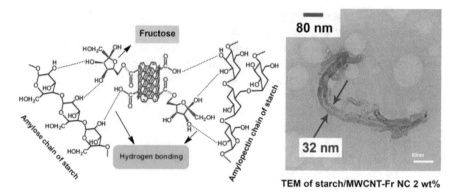

Fig. 5.3 Structure of the fructose functionalized-MWCNTs as a filler for starch films (left) and TEM micrograph of starch/MWCNT–fructose (2 wt%) nanocomposite film (right). Reprinted with permission from Ref. [65]. Copyright 2017 Elsevier

nal modified-starch hydrogen bonding, loss of short range molecular conformation, sharp increase of the overall crystallinity and larger size of micro-ordered regions of the films [62].

Other examples of original exploitations of starch and CNTs include the works describing the use of MWCNTs functionalized with low cost natural molecules, such as vitamin C or fructose, to develop films for removal of dye pollutants from wastewater [64, 65], or functionalized with glucose for the design of drug delivery systems [66]. The functionalization of MWCNTs with vitamin C (ascorbic acid) yielded a nanofiller that after incorporation into plasticized starch films, originated effective adsorbent materials for the uptake of methyl orange dye from aqueous solution [64]. Furthermore, the intrinsic biocompatibility and biodegradability, good electrical conductivity, thermal and mechanical properties, transformed these starch/MWCNTs-vitamin C films into multifunctional hybrid materials.

MWCNTs can also be functionalized with fructose, as illustrated in Fig. 5.3, and their inclusion into plasticized starch films resulted in hybrid materials with improved dispersion and compatibility between the nanofiller (0.5, 1.0 and 2.0 wt%) and the matrix [65]. Despite the absence of a proof-of-concept for the applicability of these materials, the authors highlight their potential for removal of dyes pollutants from wastewater.

Equally interesting is the functionalization of MWCNTs with glucose to develop plasticized starch-based films for application in drug delivery [66]. The starch films reinforced with MWCNTs-glucose (0.5, 1 and 2 wt%) were further reacted with oleic acid to obtain amphiphilic starch esters, which can then be used to prepare drug-loaded nanoparticles. Zolpidem, a hydrophobic model drug, was loaded into the nanoparticles and the results of entrapment efficiency, loading capacity and in vitro release tests confirmed the potential of the starch/MWCNTs-glucose hybrids for drug delivery applications [66].

5.5 Other Polysaccharides/CNTs Hybrid Materials

The use of CNTs as carbonaceous nanofillers with other polysaccharides such as ALG [69–73], carrageenan [74, 75] and hyaluronan [76, 77], has also been investigated, although to a limited extent (Table 5.1). For example, Zhao et al. [69] developed a hybrid paper based on oxidized SWCNTs, silver nanoparticles (AgNPs) and ALG. These folded structured SWCNTs hybrid papers were prepared by freeze drying followed by mechanical pressing technique and displayed excellent resistance-strain stability under various deformations, as well as good electrical conductivity. Hence, these highly bendable, foldable and conductive alginate/SWCNTs based hybrid papers are particularly suitable for applications in flexible conductors, phase change materials and temperature-driven switches [69].

Hybrid materials in the form of hydrogels can also be produced by exploiting the partnership between ALG and CNTs. As reported by Joddar et al. [70], porous and biocompatible hybrid hydrogels were fabricated through the binding of oxidized MWCNTs to ALG. Since the MWCNTs–COOH were used as the reinforcing phase within ALG, hybrid materials with enhanced mechanical and viscoelastic properties were produced. Furthermore, the cell adhesion, migration and proliferation assays confirmed the potential of these hydrogel hybrids as substrates for cell culture applications, cell therapy and tissue engineering [70].

Along the same idea, hydrogels based on ALG and CNTs were recently fabricated by Mottet et al. [71] with the purpose of developing a conductive hybrid material for probing microbial electroactivity upon application to microbial fuel cells. The hybrid hydrogels were prepared in the form of hollow spheres (beads or capsules) by homogenously mixing CNTS and ALG via surfactant assisted dispersion followed by a desorption step that engenders electrical conductivity. These conductive hybrid hydrogels are compatible with the exoelectrogenic bacteria *Geobacter sulfurreducens*, which is one of the best candidates for microbial fuel cells [71].

Meng et al. [73] developed golden coated SWCNTs nanohybrids by using the layer-by-layer self-assembly of two oppositely charged polysaccharides, namely ALG and CH, on SWNTs as bridge, followed by seed growth of gold nanoparticles (AuNPs), as illustrated in Fig. 5.4. These non-cytotoxic nanohybrids with enhanced NIR (near-infrared) absorption and HeLa cell internalization can rapidly cause localized hyperthermia, triggering cell death, and hence act as an effective photothermal converter for cancer ablation [73].

Carrageenan is another interesting polysaccharide that can be combined with CNTs for application as carriers for remotely activated drug delivery [74] and adsorbent for the removal of cationic dyes from wastewaters [75]. The first study reports the preparation of MWCNTs/carrageenan nanocomposites by blending the components, *i.e.* carrageenan, oxidized MWCNTs and a model drug (methylene blue), followed by carrageenan gelation with KCl to obtain hydrogels. The MWCNTs were used as multifunctional fillers to enhance the mechanical properties and to confer light responsive characteristics to the hydrogels. The use of an external stimuli (*e.g.*, temperature or NIR irradiation) induced the release of methylene blue from the hydrogels

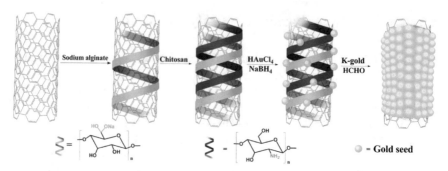

Fig. 5.4 Preparation process of golden coated SWNTs by seed growth of gold nanoparticles on the bilayer polysaccharides functionalized SWNTs. Reprinted with permission from Ref. [73]. Copyright 2014 American Chemical Society

since the MWCNTs raised the local temperature of the gel via the photothermal conversion of MWCNTs. According to the authors, these results translate into materials with potential for remotely controlled light activated drug delivery systems [74].

The second study describes the functionalization of oxidized MWCNTs with carrageenan, followed by the in situ synthesis of Fe_3O_4 nanoparticles to obtain magnetic MWCNTs/carrageenan/Fe_3O_4 nanocomposite hybrid materials for the removal of methylene blue from aqueous solution. The adsorption kinetics described by the pseudo second-order kinetic model and the adsorption isotherm data, fitted to the Langmuir isotherm model, pointed out the potential of these nanohybrids for application as magnetic adsorbents to remove cationic dyes from wastewaters [75].

Hybrids materials incorporating hyaluronan and functionalized CNTs have also been developed for application as scaffolds for tissue engineering [76] or as immunosensors for hepatitis B [77]. According to Arnal-Pastor et al. [76], scaffolds based on hyaluronan and CNTs functionalized with –COOH groups (mass fractions up to 0.05) were prepared via a two-step freeze-drying procedure and showed a highly porous network with interconnected pores of 100 to 300 μm in diameter. The presence of CNTs governed simultaneously the water sorption, porosity and mechanical properties of the scaffolds, which translated into hybrid scaffolds with customizable features to be used as inducers of neural regeneration in tissues of the central or peripheral nervous system [76].

In the other study, the fabrication of a nanohybrid electrochemical immunosensor based on hyaluronan and MWCNTs functionalized with amino groups (MWCNTs–NH_2) is described [77]. The response of the immunosensor towards the antibodies to hepatitis B core protein (anti-HBc) was linear in concentrations up to 6 ng mL^{-1} and with a detection limit of 0.03 ng mL^{-1}. These results are consistent with clinical levels and thus the hyaluronan/MWCNTs nanohybrid system can be used as a sensing platform to detect the anti-HBc [77].

References

1. De Volder MFL, Tawfick SH, Baughman RH, Hart AJ. Carbon nanotubes: present and future commercial applications. Science. 2013;339:535–9.
2. Kumar S, Nehra M, Kedia D, Dilbaghi N, Tankeshwar K, Kim KH. Carbon nanotubes: a potential material for energy conversion and storage. Prog Energy Combust Sci. 2018;64:219–53.
3. Chinnappan A, Baskar C, Kim H, Ramakrishna S. Carbon nanotube hybrid nanostructures: future generation conducting materials. J Mater Chem A. 2016;4:9347–61.
4. Iijima S, Ichihashi T. Single-shell carbon nanotubes of 1-nm diameter. Nature. 1993;363:603–5.
5. Iijima S. Helical microtubules of graphitic carbon. Nature. 1991;354:56–8.
6. Gupta S, Murthy CN, Prabha CR. Recent advances in carbon nanotube based electrochemical biosensors. Int J Biol Macromol. 2018;108:687–703.
7. Mittal G, Dhand V, Rhee KY, Park SJ, Lee WR. A review on carbon nanotubes and graphene as fillers in reinforced polymer nanocomposites. J Ind Eng Chem. 2015;21:11–25.
8. Baig Z, Mamat O, Mustapha M. Recent progress on the dispersion and the strengthening effect of carbon nanotubes and graphene-reinforced metal nanocomposites: a review. Crit Rev Solid State Mater Sci. 2018;43:1–46.
9. Chen Z, Zhang A, Wang X, Zhu J, Fan Y, Yu H, Yang Z. The advances of carbon nanotubes in cancer diagnostics and therapeutics. J. Nanomater. 2017; (3418932).
10. Pop E, Mann D, Wang Q, Goodson K, Dai H. Thermal conductance of an individual single-wall carbon nanotube above room temperature. Nano Lett. 2006;6:96–100.
11. Kumar S, Rani R, Dilbaghi N, Tankeshwar K, Kim K-H. Carbon nanotubes: a novel material for multifaceted applications in human healthcare. Chem Soc Rev. 2017;46:158–96.
12. Vivekanandhan S, Schreiber M, Muthuramkumar S, Misra M, Mohanty AK. Carbon nanotubes from renewable feedstocks: a move toward sustainable nanofabrication. J Appl Polym Sci. 2017;134:44255.
13. Research and Markets. https://www.researchandmarkets.com/reports/4343271/carbon-nanotubes-cnt-market-by-type-single#relb0 (2018). Accessed 28 Mar 2018.
14. Xu J, Cao Z, Zhang Y, Yuan Z, Lou Z, Xu X, Wang X. A review of functionalized carbon nanotubes and graphene for heavy metal adsorption from water: preparation, application, and mechanism. Chemosphere. 2018;195:351–64.
15. Rashid MH-O, Ralph SF. Carbon nanotube membranes: synthesis, properties, and future filtration applications. Nanomaterials. 2017;7:99.
16. Pardo J, Peng Z, Leblanc RM. Cancer targeting and drug delivery using carbon-based quantum dots and nanotubes. Molecules. 2018;23:378.
17. Yengejeh SI, Kazemi SA, Öchsner A. Carbon nanotubes as reinforcement in composites: a review of the analytical, numerical and experimental approaches. Comput Mater Sci. 2017;136:85–101.
18. Ghoshal S. Polymer/Carbon Nanotubes (CNT) Nanocomposites processing using additive manufacturing (three-dimensional printing) technique: an overview. Fibers. 2017;5:40.
19. Reales OAM, Toledo RD. Filho, A review on the chemical, mechanical and microstructural characterization of carbon nanotubes-cement based composites. Constr Build Mater. 2017;154:697–710.
20. Qi H, Mäder E, Liu J. Electrically conductive aerogels composed of cellulose and carbon nanotubes. J. Mater. Chem. A. 2013;1:9714–20.
21. Qi H, Mäder E, Liu J. Unique water sensors based on carbon nanotube-cellulose composites. Sens Actuators B. 2013;185:225–30.
22. Pang Z, Sun X, Wu X, Nie Y, Liu Z, Yue L. Fabrication and application of carbon nanotubes/cellulose composite paper. Vacuum. 2015;122:135–42.
23. Ahmad AL, Jawad ZA, Low SC, Zein SHS. A cellulose acetate/multi-walled carbon nanotube mixed matrix membrane for CO_2/N_2 separation. J Memb Sci. 2014;451:55–66.
24. Casella IG, Gioia D, Rutilo M. A multi-walled carbon nanotubes/cellulose acetate composite electrode (MWCNT/CA) as sensing probe for the amperometric determination of some catecholamines. Sens Actuators B. 2018;255:3533–40.

25. Mandal B, Das D, Rameshbabu AP, Dhara S, Pal S. A biodegradable, biocompatible transdermal device derived from carboxymethyl cellulose and multi-walled carbon nanotubes for sustained release of diclofenac sodium. RSC Adv. 2016;6:19605–11.
26. Soheilmoghaddam M, Adelnia H, Sharifzadeh G, Wahit MU, Wong TW, Yussuf AA. Bio-nanocomposite regenerated cellulose/single-walled carbon nanotube films prepared using ionic liquid solvent. Cellulose. 2017;24:811–22.
27. Cheng F, Liu C, Li H, Wei X, Yan T, Wang Y, Song Y, He J, Huang Y. Carbon nanotube-modified oxidized regenerated cellulose gauzes for hemostatic applications. Carbohydr Polym. 2018;183:246–53.
28. Zeng X, Deng L, Yao Y, Sun R, Xu J, Wong C-P. Flexible dielectric papers based on biodegradable cellulose nanofibers and carbon nanotubes for dielectric energy storage. J Mater Chem C. 2016;4:6037–44.
29. Yamakawa A, Suzuki S, Oku T, Enomoto K, Ikeda M, Rodrigue J, Tateiwa K, Terada Y, Yano H, Kitamura S. Nanostructure and physical properties of cellulose nanofiber-carbon nanotube composite films. Carbohydr Polym. 2017;171:129–35.
30. Sun J, Zhang C, Yuan Z, Ji X, Fu Y, Li H, Qin M. Composite films with ordered carbon nanotubes and cellulose nanocrystals. J Phys Chem C. 2017;121:8976–81.
31. Bai L, Bossa N, Qu F, Winglee J, Li G, Sun K, Liang H, Wiesner MR. Comparison of hydrophilicity and mechanical properties of nanocomposite membranes with cellulose nanocrystals and carbon nanotubes. Environ Sci Technol. 2017;51:253–62.
32. Aritonang HF, Kamu VS, Ciptati C, Onggo D, Radiman CL. Performance of platinum nanoparticles/multiwalled carbon nanotubes/bacterial cellulose composite as anode catalyst for proton exchange membrane fuel cells. Bull Chem React Eng Catal. 2017;12:287–92.
33. Gutiérrez-Hernández JM, Escobar-García DM, Escalante A, Flores H, González FJ, Gatenholm P, Toriz G. In vitro evaluation of osteoblastic cells on bacterial cellulose modified with multi-walled carbon nanotubes as scaffold for bone regeneration. Mater Sci Eng C. 2017;75:445–53.
34. Meng Q, Manas-Zloczower I. Carbon nanotubes enhanced cellulose nanocrystals films with tailorable electrical conductivity. Compos Sci Technol. 2015;120:1–8.
35. El Badawi N, Ramadan AR, Esawi AMK, El-Morsi M. Novel carbon nanotube-cellulose acetate nanocomposite membranes for water filtration applications. Desalination. 2014;344:79–85.
36. Qi H, Liu J, Pionteck J, Pötschke P, Mäder E. Carbon nanotube-cellulose composite aerogels for vapour sensing. Sens Actuators B. 2015;213:20–6.
37. Qi H, Schulz B, Vad T, Liu J, Mäder E, Seide G, Gries T. Novel carbon nanotube/cellulose composite fibers as multifunctional materials. ACS Appl Mater Interfaces. 2015;7:22404–12.
38. Dichiara AB, Song A, Goodman SM, He D, Bai J. Smart papers comprising carbon nanotubes and cellulose microfibers for multifunctional sensing applications. J Mater Chem A. 2017;5:20161–9.
39. Kuzmenko V, Naboka O, Haque M, Staaf H, Göransson G, Gatenholm P, Enoksson P. Sustainable carbon nanofibers/nanotubes composites from cellulose as electrodes for supercapacitors. Energy. 2015;90:1490–6.
40. Garcia I, Azcune I, Casuso P, Carrasco PM, Grande HJ, Cabañero G, Katsigiannopoulos D, Grana E, Dimos K, Karakassides MA, Odriozola I, Avgeropoulos A. Carbon nanotubes/chitin nanowhiskers aerogel achieved by quaternization-induced gelation. J Appl Polym Sci. 2015;132:42547.
41. Chen C, Yang C, Li S, Li D. A three-dimensionally chitin nanofiber/carbon nanotube hydrogel network for foldable conductive paper. Carbohydr Polym. 2015;134:309–13.
42. Wu S, Duan B, Lu A, Wang Y, Ye Q, Zhang L. Biocompatible chitin/carbon nanotubes composite hydrogels as neuronal growth substrates. Carbohydr Polym. 2017;174:830–40.
43. Wu S, Duan B, Zeng X, Lu A, Xu X, Wang Y, Ye Q, Zhang L. Construction of blood compatible lysine-immobilized chitin/carbon nanotube microspheres and potential applications for blood purified therapy. J Mater Chem B. 2017;5:2952–63.
44. Salam MA, El-Shishtawy RM, Obaid AY. Synthesis of magnetic multi-walled carbon nanotubes/magnetite/chitin magnetic nanocomposite for the removal of Rose Bengal from real and model solution. J Ind Eng Chem. 2014;20:3559–67.

45. Singh N, Chen J, Koziol KK, Hallam KR, Janas D, Patil AJ, Strachan A, Hanley JG, Rahatekar SS. Chitin and carbon nanotube composites as biocompatible scaffolds for neuron growth. Nanoscale. 2016;8:8288–99.
46. Xu H, Wang L, Luo J, Song Y, Liu J, Zhang S, Cai X. Selective recognition of 5-hydroxytryptamine and dopamine on a multi-walled carbon nanotube-chitosan hybrid film-modified microelectrode array. Sensors. 2015;15:1008–21.
47. Zhang Q, Qing Y, Huang X, Li C, Xue J. Synthesis of single-walled carbon nanotubes–chitosan nanocomposites for the development of an electrochemical biosensor for serum leptin detection. Mater Lett. 2018;211:348–51.
48. Shukla SK, Lavon A, Shmulevich O, Ben-Yoav H. The effect of loading carbon nanotubes onto chitosan films on electrochemical dopamine sensing in the presence of biological interference. Talanta. 2018;181:57–64.
49. Ouyang A, Gong Q, Liang J. Carbon nanotube-chitosan composite beads with radially aligned channels and nanotube-exposed walls for bilirubin adsorption. Adv Eng Mater. 2015;17:460–6.
50. Liu X, Zhang Y, Ma D, Tang H, Tan L, Xie Q, Yao S. Biocompatible multi-walled carbon nanotube-chitosan-folic acid nanoparticle hybrids as GFP gene delivery materials. Colloids Surf B Biointerfaces. 2013;111:224–31.
51. Patel KD, Kim TH, Lee EJ, Han CM, Lee JY, Singh RK, Kim HW. Nanostructured biointerfacing of metals with carbon nanotube/chitosan hybrids by electrodeposition for cell stimulation and therapeutics delivery. ACS Appl Mater Interfaces. 2014;6:20214–24.
52. Popuri SR, Frederick R, Chang C-Y, Fang S-S, Wang C-C, Lee L-C. Removal of copper (II) ions from aqueous solutions onto chitosan/carbon nanotubes composite sorbent. Desalin Water Treat. 2014;52:691–701.
53. Liu H, Gong C, Wang J, Liu X, Liu H, Cheng F, Wang G, Zheng G, Qin C, Wen S. Chitosan/silica coated carbon nanotubes composite proton exchange membranes for fuel cell applications. Carbohydr Polym. 2016;136:1379–85.
54. Ou Y, Tsen W-C, Gong C, Wang J, Liu H, Zheng G, Qin C, Wen S. Chitosan-based composite membranes containing chitosan-coated carbon nanotubes for polymer electrolyte membranes. Polym Adv Technol. 2018;29:612–22.
55. Wang J, Gong C, Wen S, Liu H, Qin C, Xiong C, Dong L. Proton exchange membrane based on chitosan and solvent-free carbon nanotube fluids for fuel cells applications. Carbohydr Polym. 2018;186:200–7.
56. Figueredo F, González-Pabón MJ, Cortón E. Low cost layer by layer construction of CNT/Chitosan flexible paper-based electrodes: a versatile electrochemical platform for point of care and point of need testing. Electroanalysis. 2018;30:497–508.
57. Seo SJ, Kim JJ, Kim JH, Lee JY, Shin US, Lee EJ, Kim HW. Enhanced mechanical properties and bone bioactivity of chitosan/silica membrane by functionalized-carbon nanotube incorporation. Compos Sci Technol. 2014;96:31–7.
58. Trigueiro JPC, Silva GG, Pereira FV, Lavall RL. Layer-by-layer assembled films of multi-walled carbon nanotubes with chitosan and cellulose nanocrystals. J Colloid Interface Sci. 2014;432:214–20.
59. Song K, Gao A, Cheng X, Xie K. Preparation of the superhydrophobic nano-hybrid membrane containing carbon nanotube based on chitosan and its antibacterial activity. Carbohydr Polym. 2015;130:381–7.
60. Zhu Y, Liu X, Yeung KWK, Chu PK, Wu S. Biofunctionalization of carbon nanotubes/chitosan hybrids on Ti implants by atom layer deposited ZnO nanostructures. Appl Surf Sci. 2017;400:14–23.
61. Morales NJ, Candal R, Famá L, Goyanes S, Rubiolo GH. Improving the physical properties of starch using a new kind of water dispersible nano-hybrid reinforcement. Carbohydr Polym. 2015;127:291–9.
62. Liu S, Li X, Chen L, Li L, Li B, Zhu J. Understanding physicochemical properties changes from multi-scale structures of starch/CNT nanocomposite films. Int J Biol Macromol. 2017;104:1330–7.

63. Shahbazi M, Rajabzadeh G, Sotoodeh S. Functional characteristics, wettability properties and cytotoxic effect of starch film incorporated with multi-walled and hydroxylated multi-walled carbon nanotubes. Int J Biol Macromol. 2017;104:597–605.
64. Mallakpour S, Rashidimoghadam S. Starch/MWCNT-vitamin C nanocomposites: Electrical, thermal properties and their utilization for removal of methyl orange. Carbohydr Polym. 2017;169:23–32.
65. Mallakpour S, Khodadadzadeh L. Fructose functionalized MWCNT as a filler for starch nanocomposites: Fabrication and characterizations. Prog Org Coat. 2018;114:244–9.
66. Mallakpour S, Khodadadzadeh L. Ultrasonic-assisted fabrication of starch/MWCNT-glucose nanocomposites for drug delivery. Ultrason Sonochem. 2018;40:402–9.
67. Cheng J, Zheng P, Zhao F, Ma X. The composites based on plasticized starch and carbon nanotubes. Int J Biol Macromol. 2013;59:13–9.
68. Swain SK, Pradhan AK, Sahu HS. Synthesis of gas barrier starch by dispersion of functionalized multiwalled carbon nanotubes. Carbohydr Polym. 2013;94:663–8.
69. Zhao S, Gao Y, Li J, Zhang G, Sun R, Wong CP. Facile preparation of folded structured single-walled carbon nanotube hybrid paper: Toward applications as flexible conductor and temperature-driven switch. Carbon. 2015;95:987–94.
70. Joddar B, Garcia E, Casas A, Stewart CM. Development of functionalized multi-walled carbon-nanotube-based alginate hydrogels for enabling biomimetic technologies. Sci Rep. 2016;6:32456.
71. Mottet L, Le Cornec D, Noël J-M, Kanoufi F, Delord B, Poulin P, Bibette J, Bremond N. Conductive hydrogel based on alginate and carbon nanotubes for probing microbial electroactivity. Soft Matter. 2018;14:1434–41.
72. Shi X, Zheng Y, Wang C, Yue L, Qiao K, Wang G, Wang L, Quan H. Dual stimulus responsive drug releasing under the interaction of pH value and pulsatile electric field for bacterial cellulose/sodium alginate/multi-walled carbon nanotubes hybrid hydrogel. RSC Adv. 2015;5:41820–9.
73. Meng L, Xia W, Liu L, Niu L, Lu Q. Golden single-walled carbon nanotubes prepared using double layer polysaccharides bridge for photothermal therapy. ACS Appl Mater Interfaces. 2014;6:4989–96.
74. Estrada AC, Daniel-da-Silva AL, Trindade T. Photothermally enhanced drug release by κ-carrageenan hydrogels reinforced with multi-walled carbon nanotubes. RSC Adv. 2013;3:10828–36.
75. Duman O, Tunç S, Polat TG, Bozoğlan BKI. Synthesis of magnetic oxidized multiwalled carbon nanotube-κ-carrageenan-Fe_3O_4 nanocomposite adsorbent and its application in cationic Methylene Blue dye adsorption. Carbohydr Polym. 2016;147:79–88.
76. Arnal-Pastor M, Tallà Ferrer C, Herrero MH, Aldaraví AM-G, Pradas MM, Vallés-Lluch A. Scaffolds based on hyaluronan and carbon nanotubes gels. J Biomater Appl. 2016;31:534–43.
77. Cabral DGA, Lima ECS, Moura P, Dutra RF. A label-free electrochemical immunosensor for hepatitis B based on hyaluronic acid-carbon nanotube hybrid film. Talanta. 2016;148:209–15.

Chapter 6
Conclusions and Future Perspectives

The evolution of ideas towards the development of environmentally friendly materials derived from renewable resources has spotlighted biopolymers (and polysaccharides in particular) as substrates to engineer multifunctional materials, as corroborated by the extent and diversity of the publications portrayed in the present book. The drive for this development culminates in the combination of polysaccharides, such as cellulose, chitin, chitosan and starch, with metal nanoparticles (NPs) (Chap. 2), metal oxide NPs (Chap. 3), graphene (Chap. 4) and carbon nanotubes (Chap. 5) for the fabrication of hybrid materials with tailorable properties and a broad spectrum of applications.

In the last 5 years, the interdisciplinary area of polysaccharide-based hybrid materials has been focussed mainly on the use of performance-property driven methodologies, as well as biomimetic or bioinspired processing approaches that favour low energy consumption with recyclable media or solventless procedures. The type of polysaccharide and inorganic component will impart distinct properties to the resulting hybrid materials, and thus will dictate the field of application. Bacterial cellulose, for example, is particularly apposite for applications requiring never-dried membranes or films with good dimensional stability and mechanical properties, while chitosan is more appropriate for applications where its film-forming ability and antimicrobial activity are important, and starch for applications where edibility and thermoplastic properties are imperative.

On the other hand, palladium NPs for instance have a huge potential for application in catalysis due to the catalytic activity of this metal, whereas the magnetic properties of iron oxides make them suitable as magneto-responsive platforms. Furthermore, graphene and graphene oxide, known for their reinforcing potential and porous nature, are appropriate for application as adsorbent materials (*e.g.*, for the removal of water contaminants), while the electrical conductivity of carbon nanotubes enables their application as electrodes.

Scalability issues, lack of environmental and biological risk assessment, and the absence of regulatory guidelines for nanomaterials are the main constraints hindering the commercial translation of most of the polysaccharide-based hybrids enumerated

© The Author(s), under exclusive license to Springer Nature Switzerland AG 2018   115
C. Vilela et al., *Polysaccharide Based Hybrid Materials*, Biobased Polymers,
https://doi.org/10.1007/978-3-030-00347-0_6

in the present book. Nevertheless, the interest of chemists, physicists, biologists and materials scientists on polysaccharide-based hybrids will continue to expand towards the design of smart hierarchical bioinspired structures with singular and tailor-made properties for application in almost all fields of modern science and technology. In fact, it is reasonable to assume that the increasing attention given by the academia, but also by the industry (although in an early stage), will drive the research on polysaccharide-based hybrids towards viable commercial materials.

Printed in the United States
By Bookmasters